T0146234

Reading Galileo

READING GALILEO

Scribal Technologies and the *Two New Sciences*

RENÉE RAPHAEL

Johns Hopkins University Press
Baltimore

© 2017 Johns Hopkins University Press
All rights reserved. Published 2017
Printed in the United States of America on acid-free paper
9 8 7 6 5 4 3 2 1

Johns Hopkins University Press
2715 North Charles Street
Baltimore, Maryland 21218-4363
www.press.jhu.edu

Library of Congress Cataloging-in-Publication Data

Names: Raphael, Renée Jennifer, 1979–
Title: Reading Galileo : scribal technologies and the Two new sciences /
    Renée Raphael.
Other titles: Two new sciences
Description: Baltimore : Johns Hopkins University Press, 2017 |
    Includes bibliographical references and index.
Identifiers: LCCN 2016019647 | ISBN 9781421421773 (hardcover : alk. paper) |
    ISBN 1421421771 (hardcover : alk. paper) | ISBN 9781421421780 (electronic) |
    ISBN 142142178X (electronic)
Subjects: LCSH: Galilei, Galileo, 1564–1642. | Galilei, Galileo, 1564–1642
    —Influence. | Galilei, Galileo, 1564–1642. Discorsi e dimostrazioni matematiche.
    English. | Science, Renaissance. | Mechanics—Early works to 1800. | Physics
    —Early works to 1800.
Classification: LCC QB36.G2 R2155 2017 | DDC 530—dc23
LC record available at https://lccn.loc.gov/2016019647

A catalog record for this book is available from the British Library.

*Special discounts are available for bulk purchases of this book. For more information,
please contact Special Sales at 410-516-6936 or specialsales@press.jhu.edu.*

Johns Hopkins University Press uses environmentally friendly book materials,
including recycled text paper that is composed of at least 30 percent post-consumer
waste, whenever possible.

CONTENTS

This book is about Galileo's debts to the scholarly traditions he inherited from his contemporaries and predecessors. Like Galileo, I owe much to the intellectual communities of which I have been a part while writing.

My first thanks go to Anthony Grafton, who supported the project from its inception, read through countless drafts, and was unfailingly encouraging throughout the process. For introducing me to Galileo's *Two New Sciences*, I thank Daniel Garber and the late Michael Mahoney, whose jointly led graduate seminar provided the opportunity to puzzle through the text. Important transformations in my approach and methodology, as well as advice and a critical eye, came from Ann Blair in multiple venues and interactions, especially the faculty seminar "Managing Scholarly Information before the Modern Age," which she led at the Folger Shakespeare Library in spring 2011. My attention to book production and images was inspired by a fruitful year as a postdoctoral associate with a project funded by the Arts and Humanities Research Council in the United Kingdom, "Diagrams, Images, and the Transformation of Astronomy, 1450–1650," led by Nicholas Jardine, Sachiko Kusukawa, and Liba Taub. It also owes much to Nick Wilding's expertise.

Multiple institutions and individuals helped the project along at various stages. I want to thank my colleagues at the University of Alabama and the University of California, Irvine, who judged my work on Galileo and mechanics a worthy addition to their departments. Holly Grout, Amy Holmes-Tagchungdarpa, David Igler, Heather Kopelson, George McClure, Nancy McLoughlin, David Michelson, Laura Mitchell, Rachel O'Toole, Allison Perlman, and Jenny Shaw were generous colleagues who helped me navigate my first years of teaching and the frustrations of the publishing process. My heartfelt appreciation goes to Bob Moeller, who read through multiple drafts and offered an open ear and sage counsel. Much of the book in its present form was researched and written at Villa I Tatti, The Harvard University Center for Italian Renaissance Studies. I want to thank all the members of the I Tatti community during the 2012–13 academic year, especially Andrew Berns, Giancarlo Casale, Emanuela Ferretti, Philippa Jackson, Evan MacCarthy, Alina Payne, Laura Refe, Jessica Richardson, Marco Sgarbi, and Nicholas Terpstra. My thanks to Lina Pertile, Anna Bensted, and Nelda Cantarella Ferace, as well as the staff at I Tatti, for a welcoming and productive year.

Many others provided support, advice, and encouragement. I thank Eileen Reeves and Michael Gordin for reading parts of the book and helping me navigate the world of twenty-first-century publishing. Karin Ekholm, Rupali Mishra, and Margaret Schotte were attentive readers at the start of the project whose continuing friendship has been invaluable. Many individuals offered key advice and support along the way, including Domenico Bertoloni Meli, Massimo Bucciantini, Jochen Büttner, Marta Fattori, Mordechai Feingold, Paula Findlen, Paolo Galluzzi, Hilary Gatti, John Heilbron, Adrian Johns, Vera Keller, Elaine Leong, Gideon Manning, William Newman, Carol Pal, Jürgen Renn, William Shea, Giorgio Stabile, Matteo Valleriani, and William Patrick Watson. Sophie Roux and Isabelle Pantin very warmly welcomed me in Paris, sharing with great kindness their own expertise and contacts. Robert Westman facilitated my introduction to the UC community and generously read the entire manuscript. Monica Azzolini was instrumental in securing many of the images in the book. Darin Hayton offered a venue to present some of my findings and serendipitously facilitated an introduction to Matthew McAdam. I want to thank Matt for his help shepherding the book into print and especially for his sensitivity to its aims and audience. The suggestions of anonymous reviewers and the editorial staff at Johns Hopkins University Press shaped and greatly improved its final form.

The approach to the *Two New Sciences* undertaken in the following pages would not have been possible without access to the rich archival and rare-book holdings of many libraries. I want to thank the staff, librarians, and archivists at the Biblioteca Nazionale Centrale di Firenze, the Biblioteca Medicea Laurenziana, the Biblioteca Casanatense, the Archivum Romanum Societatis Iesu, the Archivo Historico de la Pontificia Universidad Gregoriana, the Biblioteca Nazionale Centrale di Roma, the Bodleian Library, the Bibliothèque de l'Institute de France, and the Bibliothèque Sainte-Geneviève. R. P. Martín María Morales and Cristina Berna, at the archive of the Gregorian University in Rome, were especially accommodating, while Alexandra Franklin welcomed me to the Bodleian during the summer of 2014. Elma

Brenner very generously verified the details of one copy of the *Two New Sciences* held at the Wellcome Library. Marie-Hélène de La Mure, at the Bibliothèque Sainte-Geneviève, corresponded with me regarding the Mersenne copy discussed in chapter 3; she, Mark Smith, and Annie Charon, through the introduction of Isabelle Pantin, very generously assisted in the initial decipherment and identification of the copy.

Research for the book was supported by a Fulbright Institute of International Education research fellowship to Italy, a National Science Foundation graduate research fellowship, a Villa I Tatti Hanna Kiel Fellowship, and a Bodleian Library Research Grant sponsored by the Renaissance Society of America. Images and permissions were subsidized through a UCI Humanities Commons publication-support grant. Parts of this book have been published in different forms elsewhere. Select case studies in chapters 4 through 6 appear in "Reading Galileo's Discorsi in the Early Modern University," *Renaissance Quarterly* 68.2 (Summer 2015): 558–96. I wish to thank the editor and publisher for permission to use this material here.

Despite the intellectual community out of which it arises, producing a book is often a lonely endeavor. I thank my parents, Marc and Claudette Raphael, for their support on the journey, as well as Julio Proaño and Gabriela, who have weathered the trials of Galileo along with me.

Reading Galileo

Introduction

In 1633, Galileo Galilei (1564–1642) was approaching seventy years of age. He had just been condemned by the Catholic Church, which had subjected him to formal imprisonment and prohibited the sale of his *Dialogue on the Two World Systems*, whose publication the preceding year had occasioned his trial before the Holy Office. Galileo was first confined to house arrest in Rome at the Florentine Embassy, but within a month Pope Urban VIII granted him permission to be removed to the custody of Archbishop Ascanio Piccolomini of Siena. Shortly thereafter, he was sent to Arcetri, outside Florence, where he remained until his death. These last years were not easy on him. In 1634 Galileo's favorite daughter, Virginia, Sister Marie Celeste, passed away. Galileo himself suffered from increasingly poor eyesight and had lost his vision permanently by 1638.[1]

Despite these personal difficulties, Galileo—or, more precisely, Galileo along with a network of correspondents, eager assistants, and students—undertook a variety of projects during this period. In 1635, through the efforts of Elia Diodati (1576–1661) in Paris, a Latin translation of the 1632 *Dialogue* was completed by the professor of history Matthias Bernegger (1582–1640) and printed in Strasbourg.[2] Through correspondence and marginal annotations added to printed books, Galileo responded to criticisms and comments on his earlier publications.[3] Vincenzio Renieri (1606–1647) assisted Galileo in his attempts to sell to the Dutch government a method for determining longitude based on the tables he had derived of the motions of the four satellites of Jupiter.[4] In discussions with Pierre Carcavy (ca. 1600–1684) in Toulouse, Diodati in Paris, and Louis Elsevier (1604–1670) in Leiden, Galileo pursued the possibility of printing a complete collection of his published and unpublished works, with his collaborators going so far as to commission new engravings for the illustrations and exploring the prospect of translating the Italian texts into Latin.[5] And in 1638 the Elsevier press, in

Leiden, saw into print what became Galileo's final published work, his *Discorsi e dimostrazioni matematiche intorno a due nuove scienze*, better known in English as the *Two New Sciences*.

The *Two New Sciences* has traditionally been described as Galileo's greatest contribution to modern science, paving the way for Isaac Newton (1642–1727) and others.[6] The work is often characterized as containing two "new" sciences: a science of the resistance of materials, which describes in mathematical terms bodies' resistance to breaking, and a science of motion, outlining the quantitative relationship between the times, distances, and speeds of bodies in free fall and projectile motion. For mid-twentieth-century scholars responsible for the grand narrative of what became known as the Scientific Revolution, the text played a transformative role. Alexandre Koyré (1892–1964), for example, considered Galileo's findings on free fall to be the first step toward the law of inertia and the instigator of a profound transformation in seventeenth-century intellectual activity that would inaugurate modern science.[7] For A. Rupert Hall (1920–2009), Galileo's *Two New Sciences* was the work that "in intellectual quality and weight . . . almost outweighs all the rest of his writings." Its importance lay in its "induction of fundamental principles which henceforth dominate[d] a whole field of science."[8] These and other historians portrayed the *Two New Sciences* as introducing key ideas that, together with contributions by René Descartes (1596–1650), were taken up and developed further by Christiaan Huygens (1629–1695) and eventually Newton.[9]

The book's contents, historians argue, coincide with the true intellectual questions that motivated Galileo before he became entangled in his defense of the Copernican system and the fame and wealth that followed his telescopic discoveries of 1609.[10] Similarly, insomuch as historians characterize the *Two New Sciences* as containing Galileo's greatest contributions to modern science, they have assumed as well that Galileo recognized and intended the work as such. In the words of one recent scholar, the *Two New Sciences* is "in some sense [Galileo's] scientific legacy" and also the representation of "his strongest claim to rigor and systematicity."[11] These descriptions suggest that the *Two New Sciences* was Galileo's tour de force, a self-standing example of his method and results.

The *Two New Sciences*, however, lacks the cohesiveness of many of Galileo's earlier publications. Written as a dialogue, it treats at least two different themes, the science of materials and the science of motion. These topics are only loosely linked through the conversation between Galileo's characters.

When they have tried to make sense of the whole, historians for the most part have done so by attempting to identify relationships between its parts, to see the text as a coherent presentation of Galileo's thought.[12]

This book began as an attempt to explore these issues by examining the responses of period readers. Like many previous historians, I was puzzled by the disparate elements contained in the *Two New Sciences*, elements that often seemed at odds with the book's status as a harbinger of modern science. Moreover, despite intensive examination of the reactions of a handful of readers, the larger history of the reception of the text remained an open question. I was also interested in the larger historiographical forces at play in shaping our narratives of early modern science and of the place of Galileo's *Two New Sciences* in them. Despite its earlier celebrated status, the 1638 work no longer occupies a pivotal place in the literature. In recent decades the study of early modern science has moved beyond a study of the mathematical sciences and their key expositors to create a more inclusive narrative with a broader range of knowledge traditions, practitioners, and approaches.[13] One reason the *Two New Sciences* is no longer central to these efforts is that new questions and methods have not been applied to it. I aimed to direct just such a new set of approaches—ones focused on readers, reception, and ordinary practitioners—to this once-prized text. I sought out readers not included in existing studies of the *Two New Sciences* and its reception to see how they evaluated the book. I wanted to determine what these readers understood when they read the text, with which parts they agreed or disagreed, and the importance they assigned to the findings it presented.

The sources, however, did not behave as I had anticipated. To be sure, I found individuals unrepresented in the historiography who read the *Two New Sciences* and wrote about it. But their reactions did not conform to my expectations. Existing studies largely explore how readers evaluated the *Two New Sciences*, emphasizing when readers agreed or disagreed with Galileo and the methods—experimentation, mathematical calculation, or philosophical reasoning—they used to arrive at their determinations.[14] The new readers I identified, in contrast, seemed to care little about such issues. None of them, for example, provided a page-by-page review of the *Two New Sciences*, as Descartes did in his letter of 11 October 1638 to Marin Mersenne (1588–1648), in which he noted the errors, omissions, and contradictions in the text.[15] Some readers remarked on the congruence (or lack of congruence) of Galileo's findings with real-world experience, but none reported having recruited students and colleagues to test Galileo's findings on motion, as did, for example, the Jesuit astronomer Giovanni Battista Riccioli (1598–1671) or the atomist Pierre Gassendi (1592–1655). Nor did these new readers follow the pattern emphasized in the historiography of such prolific Jesuit commentators as Niccolò Cabeo (1586–1650) and Honoré Fabri (1608–88), who picked apart the logic underlying Galileo's science of motion to assert that whether Galileo's science agreed with experiment was immaterial, for it was based on faulty conceptions of the continuum.

The reason these readers defied my expectations, I came to conclude, was that they read the *Two New Sciences* using different methods and with distinct goals from our own. Such a conclusion should come as no surprise to historians of the book and of reading, who have long emphasized the mutability of texts and the historical specificity of reading and writing practices. This conclusion, however, was long in coming for me, because it did not conform to the established narrative regarding the reception of the *Two New Sciences*. It also sat uneasily with the notion that establishing validity is an intrinsic and timeless aspect of doing science.

My Scholarly Methods

In order to make sense of the reactions of Galileo's readers, both new and long established, this book takes its lead from three currents in scholarship on early modern intellectual history. It draws, first, on the methods of historians of the book and its readers. The importance of the history of the book to the history of science was proclaimed by Elizabeth Eisenstein, who tantalizingly declared that the standardization, dissemination, and fixity of print made transformations in early modern science possible. The counter to this argument—that standardization, dissemination, and fixity were not inherent in print but developed alongside these intellectual changes—has prompted scholars to consider in tandem the production of knowledge and the production of texts.[16] Methodologically, I draw on the approach employed by Lisa Jardine and Anthony Grafton in their study of Gabriel Harvey's (1545–1630) reading of the Roman historian Livy. As part of their argument for the close connection between classical texts and Tudor political practice, Jardine and Grafton proposed a transactional model of reading in which a single text gives rise to a plurality of possible responses. They demonstrated the value of close examination of an individual's traces of reading and the circumstances that prompted it, emphasized the multiplicity of methods individuals applied to the reading of a given text, and argued that early modern reading was often oriented for specific, practical ends that were not immediately obvious from the printed text.[17]

The second area on which this book draws is the burgeoning study of early modern Aristotelianism. A number of historians have demonstrated the continuing vitality of the Aristotelian tradition through the end of the seventeenth century, emphasizing its flexibility in responding to novel approaches and its influence on the thought of period innovators.[18] Key to this direction of scholarship have been the insights of Charles Schmitt, who argued that early modern Aristotelianism was both multivalent and eclectic.[19] According to Schmitt, we should think in terms of Renaissance Aristotelianisms, as opposed to a singular Renaissance Aristotelianism. This Aristotelianism(s), moreover, was an open tradition, one that since antiquity had drawn freely upon more than one intellectual tradition to smooth over inconsistencies and obviate problems. Like Schmitt and the historians who succeeded him, I argue for the vitality of the Aristotelian tradition and the role its ideas and associated practices played in shaping the reception of Galileo's *Two New Sciences*.

Following Schmitt, I define eclecticism as an approach that is in dialogue with multiple intellectual traditions. Eclectic philosophical systems, like Aristotelianism or Platonism, are based on a defined corpus of "Aristotelian" or "Platonic" writings, but they openly engage with writings and approaches outside it. In the case of Galileo's *Two New Sciences*, however, the norm against which eclecticism is defined is more ambiguous, delineated partly by Galileo's own proclamations of what his approach involved and partly by how historians have characterized his approach and that of the period's New Science more generally. I thus use the term *eclectic* in two senses. First, it refers to the application of reading strategies drawn from a variety of disciplinary traditions, such as mathematics and natural philosophy. I also use the word to describe the application of traditional textual methods, such as those drawn from the Aristotelian tradition, to Galileo's text. Such application is eclectic in the sense that it defies both Galileo's rhetoric regarding the proper way to investigate nature—using experiment and mathematics, rather than texts—and historians' tendency to identify Galileo's self-identified practices with the *Two New Sciences*. Over the course of the coming chapters, it will become clear that Galileo's rhetorical proclamations of his method differed from his actual approaches. Period readers' practices, moreover, suggest that historians' characterizations of a Galilean approach to science require revision. One of my main arguments is that eclecticism—an approach that drew on both established practices and new methods—was an intellectually rigorous stance and the prevalent mode in which readers made sense of Galileo's novel claims about the natural world. Readers of the *Two New Sciences*, all of whom were trained in some capacity in the Aristotelian tradition, necessarily read his text eclectically, applying a variety of reading methods to it and appropriating only parts of Galileo's claims and approaches in their own work.

In addition to these insights, I rely, finally, on the turn in the history of science to the study of scientific practices.[20] The reading and writing of texts was perhaps the most fundamental activity of all those engaged in the study of the natural world during the period, one central to the development of arguments about the natural world and the subsequent transmission and teaching of them.[21] Reading and its related practices of note-taking and annotation are activities that have changed over time in response to texts' materialities, economic factors, and readers' goals. Readers of Greek, Latin, and other dialects in the ancient and late antique Mediterranean, for example, read aloud from texts written in *scriptura continua*, a type of writing produced without spaces or punctuation between words.[22] In the early modern period, in contrast, humanists promoted active reading, which involved coming to texts with pen in hand and devoting particular attention to faithful citation and description.[23] These note-taking tendencies were encouraged by cultural, economic, and technological factors, especially the availability of cheap paper and a period exuberance for accumulating information.[24]

Ann Blair has argued that Renaissance natural philosophy was a bookish enterprise shaped by Scholastic and humanist practices.[25] An exercise in exegesis that involved the extraction and reuse of material from texts, it had as its goal not the discovery of new ideas but the systematization of an existing body of knowledge, whose parameters were set by the writings of Aristotle and his commentators. Building on this notion of bookishness, a number of scholars have explored the way paper technologies—reading, note-taking, annotating, copying, exchanging, and the organizing of books and papers—shaped early modern knowledge production.[26] Underlying this direction of investigation is the insight that knowing more about early modern practices of reading can serve as a window into early modern cognitive practices, including observation, the art of memory, and the solidification and erosion of belief.[27]

The study of these paper tools, however, has tended to concentrate on general rules and methods. Ann Blair, for example, has focused primarily on ordinary scholars whose work was not intended to be innovative but who are useful subjects of study precisely because they aimed to instruct and describe commonly employed working methods. Jean Bodin's 1596 *Theatrum* offered, in her words, "a fine point of departure from which to examine traditional natural philosophy at its height in the late Renaissance."[28] Others engaged in a study of scholarly practices also have often focused on individuals whose abundant surviving notes provide a window into period working methods but whose scholarly output was not particularly noteworthy. William Sherman described the English lawyer Sir Julius Caesar (1558–1636) as a "remarkable reader" in his study of the marginalia left by English Renaissance readers, not for the originality of his thinking or writing but because of the "staggeringly comprehensive digest of reading notes" that he assembled over six decades, comprising more than 1,200 dense pages of notes under 1,450 headings.[29]

In other contexts, historians have shown that many seventeenth-century innovators similarly made text and scribal technologies central to their working methods. Newton revealed his own passion for paper tools in the vast sea of materials written by him that the Newton Project has made available.[30] Studies of Newton's library have made clear that he was also an extensive note-taker and annotator.[31] He took notes in the margins of his books, on flyleaves or on endpapers, on loose inserted slips of paper left in books, and in separate notebooks. He bought his own books, often making careful records of these purchases in his notebooks, exchanged books with friends, and wrote about them in his correspondence.[32] Newton's contemporary Robert Boyle (1627–1691) believed that experimentalism was central to the production of new knowledge about the natural world, yet he also made extensive use of paper technologies in his alchemical and philosophical work.[33] Across the Atlantic and then in London, George Starkey (1628–1665) vocally criticized the methods and content of traditional learning to which he had been exposed at Harvard, yet he employed empirical methods that were inextricably bound with the practices of traditional university education. Starkey's surviving notebooks reveal his continual engagement with and correction of textual sources. His empirical practices and the notes in which he recorded them, moreover, were shaped by many of the paper technologies central to his university education at Harvard, including the formalized structure of Scholastic disputations and dichotomy charts associated with Ramist pedagogy.[34]

The examples of Newton, Boyle, and Starkey reveal the extent to which self-conscious innovators employed textual practices in their studies of the natural world. This book takes on a complementary project. It examines the scholarly methods used by period readers in response to a text portrayed explicitly by its author as being novel. In many ways, it builds on previous work examining the reception of the writings of seventeenth-century innovators, including Copernicus, Descartes, Newton, and Galileo himself.[35] This contribution differs from these earlier studies, however, in the range of evidence it employs—from marginalia to printed treatises to teaching manuscripts—and in its reliance on the now substantially developed literature on period scholarly practices. I approach the question of Galileo's reception and related issues of intellectual transformation with the tools and approaches of historians of scholarly methods. Readers' practices are interpreted as indicators of the goals they brought with them to the *Two New Sciences* and the relationship they saw between their projects and Galileo's work.

The uncovering of scholarly practices in marginal annotations, notes, and subsequent publications by necessity involves a close, intensive examination of a confined group of sources. What follows is not a mere catalog of instances in which readers mentioned Galileo's *Two New Sciences*. Instead it is an interrogation both of the methods readers used to read and interpret the text and of the ways they conceptualized and evaluated passages in relation to their own scholarship. Access to the methods and activity of reading, however, is exceedingly difficult. In the words of Robert Darnton, "It should be possible to develop a history as well as a theory of reader response. Possible, but not easy; for the documents rarely show readers at work, fashioning meaning from texts, and the documents are texts themselves, which also require interpretation. Few of them are rich enough to provide even indirect access to the cognitive and affective elements of reading."[36] The historical record at times allows for close access to readers' responses to the *Two New Sciences*. Such is the case in chapters 1 through 4, which analyze richly annotated surviving copies of Galileo's text. In other instances, however, I rely on indirect evidence, such as printed texts and surviving notes, to reconstruct the practices of some of Galileo's readers who, at least according to the historical record, did not leave behind such copious annotations. To get a sense of how these reading methods and evaluations of Galileo were disseminated more broadly, I also rely on the teaching notes, textbooks, and reading notes of university professors. As is emphasized in the relevant chapters, the traditional university exercise of the lecture, or *lectio*, was, translated literally, a "reading," and much university teaching involved the reading and exposition of key texts. Teaching provides a window into how professors read the *Two New Sciences*, even when no annotations or other direct evidence of reading survive, and how they modeled this practice of reading for their students. Throughout, I rely on comparisons with other studies of period reading, drawn from a variety of genres, to put the practices of Galileo's readers in context.

How Historians Have Read the *Two New Sciences*

The canonical status of the *Two New Sciences* and the enthusiasm its results on local motion generated have prompted a tradition of scholarship focused on the genesis of its content. One question of interest has been the conceptual and mathematical structure underlying the *Two New Sciences*, especially its sections on local motion. Using the printed text, surviving notes, and Galileo's earlier treatises on motion, historians have investigated the origins and evolution of key terms and proofs describing Galileo's quantitative rules governing motion, his search for a solid methodological foundation for his new science, the relationship between Galileo's early thoughts on motion and Copernicanism, and his conceptual contributions to the development of classical mechanics.[37] Another group of scholars have focused on Galileo's reliance on experiment, emphasizing working methods over philosophical principles and attempting to make sense of Galileo's claims by replicating his experiments.[38]

Efforts also have been made to recover the knowledge Galileo shared with his contemporaries and to recognize their role in shaping Galileo's approaches and ideas.[39] Matteo Valleriani, for example, has argued that Galileo's contributions in the *Two New Sciences* grew out of his work as an artist-engineer and that these practical pursuits were directed by theoretical concerns and problems derived from Aristotelian natural philosophy.[40]

Galileo's relationship with contemporary Aristotelianism has been subjected to especially close study and at times has occasioned lively debate. Historians writing in the traditional grand narrative of the period portrayed Galileo's achievements as the result of his break with the Scholasticism of the medieval university.[41] Later historians insisted on a similar rupture, even when they allowed that other factors besides a new way of seeing precipitated it.[42] In their accounts, Galileo's education in the Scholastic university community and his continuing relationship with its institutions and members are minimized, while his embrace of novel instruments, methods, and sites of learning are stressed.

At the opposite end of the spectrum are the efforts of some historians of what we might call the continuist tradition. This line of interpretation had its beginnings with Pierre Duhem, who saw Galileo's triumph over Scholastic Aristotelianism as the culmination of a conflict that had begun in the fourteenth century. Subsequent scholars argued for continuities between medieval and early modern scholarship, emphasizing the strength and resilience of Renaissance Aristotelianism and uncovering through patient archival work the robust links between Galileo and the Scholastic commentary traditions at Pisa and the Jesuit Collegio Romano.[43] This tradition of historiography admits Galileo's novelty while emphasizing his debt to medieval Aristotelianism, stressing his youthful engagement with Aristotelian natural philosophy and his continuing reliance on logical argumentation, forms of rhetoric, and concepts drawn from the Scholastic tradition.

This book approaches the *Two New Sciences* from a different angle, asking not how Galileo developed his ideas but instead how these ideas were read and interpreted by period readers. Historical context and Galileo's relationship with the Aristotelian tradition figure prominently in the analysis, but I look to Galileo's readers and their practices, rather than my own interpretations of the writings of Galileo and his predecessors, to answer them. The question that motivates the analysis is not, how did Galileo come up with his ideas? but rather, what did Galileo's ideas and approach signify for readers of the seventeenth century? This latter question, I argue, is central to understanding the profound transformations of early modern science. Innovations in method and ideas like those Galileo presented in the *Two New Sciences* were undoubtedly vital in bringing about scientific change, but these novelties only became transformative when they were read and appropriated by a wider community of readers. This book offers a glimpse into this process of transformation by examining how one innovative text was read, used, and taught by diverse communities of readers. In a field dominated by accounts of the genesis of novel ideas and approaches, it tackles a question that has been largely ignored: how were innovative scientific ideas and methods read and appropriated to effect intellectual change?

How Galileo Wrote the *Two New Sciences*

It is not surprising that historians have focused their efforts on uncovering the genesis of Galileo's novelty, for Galileo vociferously championed his own embrace of new methodologies and conclusions. A constant refrain in Galileo's writings is the opposition between the textual approaches employed by his detractors and his own experimental and mathematical practices. In his 1623 *Assayer*, for example, he accused his opponent, the Jesuit Orazio Grassi, of thinking "that philosophy is the creation of a man, a book like the *Iliad* or *Orlando Furioso*." Galileo scolded Grassi, advising him instead to recognize that "philosophy is written in this all-encompassing book that is constantly open before our eyes, that is the universe; it is written in mathematical language."[44] In an earlier letter to Johannes Kepler, Galileo similarly criticized contemporaries who believed "that truth is to be sought not in the world or in nature but in the comparison of texts."[45] Galileo's emphasis on using nature, not the printed text, as the ultimate authority is a reproach of the bookish practices of Renaissance natural philosophy.

Galileo's rhetoric, however, belies a reliance on textual methods, many of them derived from established sixteenth-century precedents. The younger Galileo was an avid student of Aristotle and Aristotelian natural philosophy, and as part of these studies, he engaged in the standard exercise of borrowing and copying the notes of others.[46] Galileo's humanist training shaped the rhetorical strategies he employed to engage his contemporaries.[47] Despite Galileo's and his students' insistence to the contrary, moreover, he owned and worked with a very large library filled with ancient, medieval, and contemporary works.[48]

The aspects of Galileo's scholarly production that have received attention as part of his scientific practices—his experimentation, his embrace of quantitative methods, and his work with instruments—also relied on textual technologies. Galileo recorded his observations and findings in the fields of astronomy and mechanics on scratch sheets, he kept records of his teaching and workshop activities, and he annotated his contemporaries' books as he drafted responses to them.[49] Many of his observations circulated in manuscript before print, and he shared his ideas not only for the purposes of publishing them but also to obtain suggestions for improvement.[50]

Galileo's efforts to revise and see the *Two New Sciences* into print reveal how scribal technologies, including the sharing of views in manuscript, collaboration through correspondence, the reading and annotating of books, and the use of such notes in writing, were part of his arsenal of scholarly methods. They reveal, too, that Galileo employed such methods to engage the concerns of contemporaries, even those followers of Aristotle whom he often professed to disdain.

When Galileo turned to the *Two New Sciences* in 1633, he already had completed what would become a key section of his work on the science of resistance of materials. Galileo tested additional propositions on his correspondents. He sent two of these propositions to his former student Niccolò Aggiunti (1600–1635) in September of 1633. Aggiunti responded by elaborating on their implications and providing a proof to one of them. Mario Guiducci (1585–1646), an astronomer in whose name Galileo published his 1619 *Discourse on the Comets*, was the recipient of yet another proposition. Guiducci wrote back to Galileo with an alternative proof to the proposition, which he, in turn, had obtained from Andrea Arrighetti (1592–1672), a student of Galileo's own student Benedetto Castelli (1578–1643). Arrighetti also independently sent Galileo a letter with his proof, and Galileo, with Arrighetti's permission, included the alternative proof in the printed edition.[51]

This exchange of manuscript material was accompanied by Galileo's own reliance on paper technologies, including note slips and annotations. When Galileo at the end of 1636 began to revise his earlier notes on projectile motion, he employed the technology of the paper slip. Galileo pasted new diagrams over old ones and annotated the text with revised demonstrations in his distinctively aged handwriting.[52] Galileo also used this technique to correct his published works. A copy of the *Dialogue on the Two World Systems* held in the DeGolyer Collection at the University of Oklahoma contains Galileo's handwritten corrections from the errata sheet, as well as an insertion, mistakenly omitted from the first edition, made on a note slip to complete the dialogue.[53]

Annotations also provided material for Galileo as he composed his text. In 1634 Fulgenzio Micanzio (1570–1654), a member of the order of Servants of Maria and Paolo Sarpi's (1552–1623) student and biographer, referred Galileo to a book titled *Philosophical Exercises*, written by Antonio Rocco (1586–1653), a priest and philosopher who had studied under Cesare Cremonini (1550–1631).[54] Rocco's *Exercises* were intended as a response to and criticism of Galileo's 1632 *Dialogue*. Galileo read Rocco's book, apparently with some emotion. Responding to Micanzio's request for him to jot down his responses to Rocco's book in the margins and on separate sheets, Galileo sent Micanzio two different sets of marginal notes.[55] Galileo's annotations then served as a basis for a large part of Day 1 of the *Two New Sciences*.[56]

Not only did Galileo rely on his annotations as he composed the *Two New Sciences* but they were considered valuable intellectual products in their own right. Upon receiving Galileo's first set of seventy-five annotations, Micanzio expressed his pleasure with Galileo's reasoning and suggested the possibility of making and publishing "a little book" of them.[57] In October of 1634 Micanzio wrote to Galileo telling him that he had taken to his villa Galileo's *Dialogue* and Rocco's book, ostensibly accompanied by the annotations Galileo had sent separately. Reading the two in tandem, he reported, had been a transformational intellectual experience. Micanzio recounted that he found Rocco's arguments unconvincing and was drawn for the first time to notions of atomism. In the letter, he urged Galileo to add to his marginalia, for he felt that "in these annotations there were stupendous things."[58] Micanzio subsequently shared Galileo's annotations with Rocco and another acquaintance, who expressed their admiration of Galileo's reasoning but remained unconvinced by his arguments.[59]

Galileo's textual practices reveal that he was eager to engage with and persuade the very type of traditional thinker whom he was disposed to criticize in print, a type that included Rocco. One way he sought to engage such traditional readers was by addressing topics central to traditional natural philosophy. When Galileo composed Day 1 of the *Two New Sciences*, for example, he took up many of the same themes he had addressed in his 1612 *Bodies in Water*. This treatise had been instigated by an earlier debate with Tuscan natural philosophers on physical questions treated by Aristotle regarding the nature of rarefaction and condensation and the causes underlying the floating and sinking of heavy bodies.[60] A similar impulse to relate his writings to those who saw themselves as participating in the Aristotelian tradition can be seen in the annotations added to the only surviving manuscript of the *Two New Sciences*. Alongside one proposition, Galileo's scribe inserted a note indicating that one should "see Biancani's error in his book *Luoghi Matematici di Aristotele* at page 177," a reference to Giuseppe Biancani's 1615 *Aristotelis Loca Mathematica*.[61] A member of the Jesuit order, one known for its commitment to Aristotelian philosophy, Biancani described his *Loca* as a compendium of the part of the Aristotelian corpus that could be treated mathematically.[62]

Galileo's use of the dialogue genre may have been a deliberate attempt to encourage reluctant readers to engage with the substance of his work. Historians have argued that Galileo appreciated the dialogue's inherent flexibility, which allowed him to fashion arguments as he saw fit rather than make them conform to established norms.[63] In the case of his earlier *Dialogue on the Two World Systems*, the genre offered Galileo a mask behind which he could explore more controversial arguments regarding Copernicanism.[64]

The format also allowed for a meandering discourse, one that permitted digressions and demonstrations on both philosophical content and method.[65] These features permitted Galileo to demonstrate for readers, even those with reservations about his methods and conclusions, how his intellectual project potentially intersected with their own.[66]

Galileo's attempts to secure a printer for the *Two New Sciences* provided further opportunities to circulate the work in manuscript and receive feedback. When Galileo approached Micanzio to inquire about publishing the work in Venice, for example, Micanzio shared parts of Days 1 and 2 with Antoine de la Ville (1596–1656), a French engineer stationed in Venice, and with Paolo Aproino (ca. 1584–1638), who had studied with Galileo at Padua. De la Ville wrote to Galileo directly, expressing his doubts about the possibility of creating a general science of the resistance of materials since the behavior of each object depended intrinsically on the specific material of which it was composed.[67] Micanzio also shared Galileo's manuscript pages with other acquaintances. He reported to Galileo that Andrea Argoli (1570–1657), a professor of mathematics at Padua, had read Galileo's pages "with extreme pleasure" but had expressed doubts about Galileo's resolution of the problem of Aristotle's wheel and his claim for the equality of a line and a point.[68] Galileo's arguments were deemed "divine" and "sublime" by the Venetian engineer Francesco Tensini and Galileo's former student Alfonso Antonini (1584–1657).[69] Giovanni Pieroni (1586–1654) exchanged letters with Galileo as he searched for a printer in Bohemia, conveying his enthusiasm for the work and offering Galileo suggestions for improvement. These laudatory comments, stressing the novelty of Galileo's arguments, convey a similar tone to the earlier reactions of Galileo's friends and acquaintances in response to copies of the *Dialogue* Galileo had gifted to them. It has been argued that such reactions from individuals not affiliated with academic institutions indicate that the *Dialogue* and thus, by extension, the *Two New Sciences* were intended for an audience of learned men of action who were not part of the university.[70] While this interpretation likely contains some truth, the following chapters reveal that members of early modern universities also saw themselves as appropriate readers of the *Two New Sciences* and studied and shared the text with students accordingly.[71]

Despite Galileo's rhetoric emphasizing his break with traditional methods of natural philosophy, his practices reveal a reliance on and appreciation for the paper tools employed widely by his contemporaries. Of course, it is not surprising that Galileo built on traditional practices. What the forthcoming chapters suggest is that Galileo's readers more closely mirrored his practices than they did his rhetoric. They tended to see the *Two New Sciences* less as a decisive break with past styles of scholarship than Galileo proclaimed it to be.

The Physical Book, Its Distribution, and Period Editions

Eventually arrangements were made with the Elsevier press in Leiden to print the text.[72] Printing of the *Two New Sciences* was completed by mid-July 1638. This first edition is a quarto book, containing a title page, 6 pages of preliminary text, 314 pages of text and in-line images, 4 pages of an alphabetical index (labeled "table of notable things"), and 2 pages of errata. A few copies exist in a different state, perhaps from the first issue of the book. Missing from these copies are the catchword on the final page of text ("TAVO-"), the index of notable things, and the pages of errata.[73] The preliminary text includes a dedicatory letter (3 pages), a printer's letter (2 pages), and a table of contents (1 page). The 314 pages of text and images contain multiple mispaginations, so that the final page number printed is 306. They are divided into four "days" of dialogue carried out in Italian by three interlocutors. In the third and fourth days, the interlocutors read aloud a Latin treatise composed by the "Academician," who never appears in the text. These four days of dialogue are followed by an appendix on centers of gravity.[74]

The first copies were on sale in Leiden that month.[75] Books arrived in Rome in December 1638, and Galileo's correspondents reported that fifty books were sold before April of the following year at a price of two scudi apiece—a fairly sizeable sum, for a well-paid official at the Vatican earned about sixteen scudi a month.[76] This rate of sales is not especially noteworthy. An earlier bestseller, the first edition of Lodovico Dolce's 1553 Italian translation of Ovid's *Metamorphoses*, whose print run was eighteen hundred copies, for example, sold out in four months.[77] The *Two New Sciences* was also distributed slowly across Europe. It only appeared in Strasbourg in March 1639 and in Venice the following month. Galileo himself received a copy nearly a year after the printing, in June 1639, and he sent copies as gifts to his correspondents Bonaventura Cavalieri (1598–1647) and Giovanni Battista Baliani (1582–1666).[78] Letters between Mersenne and Descartes indicate that while Mersenne had managed to see the *Two New Sciences* shortly after—or even before—it was printed, in June 1638, Descartes had difficulty gaining access to the book until later that year. One surviving copy annotated by Mersenne (discussed in chapter 3), furthermore, indicates that Mersenne had managed to acquire one of the copies of the first issue of the book. By October, however, Descartes and other of Mersenne's correspondents had obtained copies.[79]

The distribution of extant copies today can provide some idea of their circulation in the seventeenth century. Thirty-three copies of the 1638 edition are held in the United Kingdom, ten in Oxford and six in Cambridge. Some of these books can be traced to seventeenth-century owners, but of course others may have been acquired by eighteenth- and nineteenth-century collectors.[80] Italian libraries hold twenty-one copies of the first edition; eight of these are in Florence, where the National Library holds copies annotated by Galileo himself and by his last surviving student, Vincenzo Viviani (1622–1703).[81] There are also nine copies in libraries in Paris, including two copies annotated by Mersenne and one owned by Kenelm Digby (1603–1665).[82] Notable exemplars of the second edition of the *Two New Sciences*, printed in the 1655–56 edition of Galileo's *Opere*, include one annotated by Christopher Wren (1632–1723) and held at the Bodleian Library at Oxford and one owned by Duke August of Brunswick-Wolfenbüttel (1579–1666).[83] The distribution of extant copies of the 1655–56 *Opere* is roughly similar to that of the 1638 edition.

The editions, translations, and extensions of the *Two New Sciences* seen into print over the course of the seventeenth century testify to the multiple meanings and genres ascribed to the text in the period. Readers and editors varied in their opinions about what was important about the book and its author. The reprint of the 1638 *Two New Sciences* found in the 1655–56 *Opere*, for example, followed the original 1638 edition faithfully with the exception of an addition to Day 3, which Galileo reportedly dictated to his student Viviani. While this reprint of the *Two New Sciences* was advertised no differently than the original edition, the title page of the *Opere* emphasized Galileo's connections to Pisa and Padua. Labeling Galileo both a "noble Florentine" and the chief philosopher and mathematician to the grand duke, the title page went on to describe Galileo as having been a reader in mathematics at the Universities of Pisa and Padua, as well as an extraordinary professor at Pisa.[84] These details were likely included as a tribute to the key role professors at these universities played in seeing the *Opere* into print.[85]

Other editions stressed Galileo's identity as a mathematician rather than his university ties or his philosophical ambitions. Less than a year after the original printing, a French summary of the book entitled *Les nouvelles pensées de Galilée* appeared in Paris. The author of this treatise was Marin Mersenne, the French Minim who was the central figure in a circle of scientific communication and correspondence that included Descartes, Gassendi, and Pierre Fermat (1601–1665).[86] Whereas the 1638 Leiden edition had identified Galileo as the "Chief Philosopher and Mathematician to the Tuscan Grand Duke," Mersenne emphasized Galileo's mathematical prowess. He labeled Galileo the duke's "Mathematician and Engineer." The book's contents were advertised as containing "marvelous inventions," "demonstrations unknown until now," and "everything most subtle relating to mechanics and physics." Mersenne singled out Galileo's findings on local motion for special mention, describing the book as treating "the proportion of movements, both natural and violent."[87] Similar treatment is found in an English translation published by Thomas Salusbury (ca. 1625–ca. 1665) that was issued as part of a two-volume work titled *Mathematical Collections and Translations*, published in 1661 and 1665.[88] Although he acknowledged Galileo's position as philosopher to the grand duke, Salusbury downplayed Galileo's philosophical ambitions and represented the *Two New Sciences* as a mathematical treatise. He titled his translation of the book *Mathematical Discourses and Demonstrations touching Two New Sciences* and emphasized in large typeface that these two new sciences touched on "Mechanicks" and "Local Motion."[89]

An alternative depiction of Galileo, linking him to his support of the Copernican hypothesis, was conveyed in a Latin translation based on the 1638 first edition and printed in 1699 in Leiden. Whereas the vernacular translations described the *Two New Sciences* as a mechanical and mathematical work, in this edition the book was represented as an addendum to the 1632 *Dialogue*. The title page begins by listing Galileo's affiliations and titles: a member of the Academy of the Lyncei, a professor at Pisa and Padua, and the "greatest philosopher and mathematician." The main part of the work is announced as Galileo's *Systema Cosmicum*, the Latin title given to the 1632 *Dialogue* in its first translation of 1635. Only at the bottom of the title page are readers informed that the publication also contains Galileo's "Treatise on Motion, never before rendered from the Italian language into Latin."[90]

In additions and extensions to the *Two New Sciences*, Galileo's students, friends, and admirers focused their efforts on particular parts of the book, emphasizing its connection to different disciplinary traditions. One such augmentation is the so-called Fifth Day, written by the ailing Galileo and his student Evangelista Torricelli (1608–1647) and published in 1674 by Viviani. Titled *Quinto Libro degli Elementi d'Euclide, ovvero Scienza universale delle proporzioni* (The fifth book of Euclid's *Elements*, or A universal science of proportions), its connection to geometry was emphasized by the decision of publishers to reprint it several times in a small two-volume Euclid. It was only added to the *Two New Sciences* in the first Florentine edition of Galileo's *Opere*, published in 1718. A sixth day, addressing the force of percussion, was elaborated by Viviani on the basis of manuscripts held by Galileo's son, Vincenzo Galilei (1606–1649), after his father's death. This sixth day was printed in the same 1718 edition as was the fifth.[91] Torricelli also sought to reformulate and extend Galileo's science of motion in his *De Motu* (On motion), which made up one part of his 1644 *Opera Geometrica*.[92] An anonymous English author combined the sections of Torricelli's *De Motu* and Galileo's *Two New Sciences* that discussed projectiles in a 1672 treatise on gunnery.[93] Other eager readers of Galileo, including Viviani, the Pisan professors of mathematics Guido Grandi (1671–1742) and Alessandro Marchetti (1633–1714), and the French engineer François Blondel (1618–1686), proposed and in some cases published treatises on the resistance of materials that were intended as extensions of Day 2 of the *Two New Sciences*.[94]

The *Two New Sciences* is not ubiquitous in period catalogs and bibliographies, though listings of the text at times do reflect the varied associations publishers, editors, and authors ascribed to it.[95] When the book was included, there was confusion regarding whether it should be categorized as a vernacular composition or grouped with other Latin texts. Readers and booksellers were not always in agreement. The London bookseller George Thomas, for example, listed the *Two New Sciences* under the heading "Italian books" in his 1647 catalog of books purchased in Italy.[96] The alternate possibility would have been to list the book in the first part of the catalog, which had no explanatory heading but contained books written in Latin in a variety of genres, including history, medicine, natural philosophy, and mechanics. The 1703 auction catalog of Robert Hooke's library, in contrast, listed the *Two New Sciences* under the heading "Latin books, etc. in Quarto," even though the catalog also included a separate section designated "Italian Books."[97]

There was also the issue of what short title to assign to the book. In certain correspondence, Mersenne referred to the *Two New Sciences* as "the books of Dialogues," a reference to the genre of the text rather than to its content.[98] Sale and auction catalogs, in contrast, identified the book by key words in the original Italian title that emphasized its mathematical and mechanical content. George Thomas, for example, listed the text as *Discorsi Mathematiche 4* (Four mathematical discourses), while the sale catalog of Hooke's library described it as *Gal. Galilei Mecanica et i Movimenti Locali* (Galileo Galilei's *Mechanics and Local Motions*). Thomas Hyde's 1674 catalog of the Bodleian Library, which was based on Thomas James's 1620 original and organized alphabetically, described the book according to an abbreviation of its printed title, *Discorsi Matematiche attenenti alla Mecanica & i Movimenti locali*, leaving out the identifiers "Demonstrations" and "Two New Sciences."[99] The 1681 catalog of Cardinal Francesco Barberini's library in Rome also abbreviated the printed title, focusing instead on the book's promised "demonstrations" and describing it as "Dimostrazioni Matematiche attenenti alla meccanica, & a i movimenti Locali."[100]

Reading the *Two New Sciences* in the Seventeenth Century

The preceding sections present a number of contradictory depictions of the *Two New Sciences*. It is a text that has been reified by generations of historians of early modern science for its contributions to modern physics, one assiduously examined for the clues it holds to the genesis and context of Galileo's thought. Galileo's own statements regarding the work emphasized its novel methodologies and rejection of traditional scholarly approaches. The process by which Galileo wrote and published the text, however, suggests that Galileo relied more heavily on traditional and collaborative scholarly practices than his rhetoric has implied. Portrayals of Galileo in later editions, as well as readers' and publishers' attempts to categorize the book, moreover, hint at the plurality of meanings and associations the *Two New Sciences* held for a seventeenth-century audience.

The following chapters take up these themes, juxtaposing the received interpretation of the *Two New Sciences* to explorations of how seventeenth-century readers interacted with the text. The first four chapters rely on surviving annotated copies of the first two editions of the *Two New Sciences*. They use these copies as a window into period readings of Galileo's text. Chapter 1, "An Anonymous Annotator, Baliani, and the 'Ideal' Reader," explores a heavily annotated copy held in the Galileo collection in Florence and annotated by an anonymous reader in Italian. Certain features of the marginalia suggest that this anonymous reader may have been Baliani, one of Galileo's correspondents who was also the author of his own treatise on local motion. The possibility that the annotations do indeed belong to Baliani provides the opportunity to explore in greater detail the received narrative of the reception of the *Two New Sciences*. I outline the primary characteristics of the expected "ideal" reader described in this historiography and then point to ways in which the annotated copy departs from this model. These departures set the agenda for the following chapters, which argue for the plurality of approaches and goals readers applied to the *Two New Sciences*.

Chapter 2, "Editing, Commenting, and Learning Math from Galileo," and chapter 3, "Modifying Authoritative Reading to New Purposes," rely on surviving annotations composed by two well-known readers of Galileo, his student Viviani and his eager promoter Mersenne. Current scholarship has slotted Viviani and Mersenne into the category of the "ideal" reader described in chapter 1, emphasizing Viviani's and Mersenne's efforts to disseminate and extend Galileo's findings. These chapters complicate this depiction by highlighting the strategies Mersenne and Viviani employed to read and annotate Galileo's text. Viviani emerges as a reader who relied on accepted sixteenth-century textual practices associated with the mixed-mathematical tradition to read, edit, annotate, and extend Galileo's work. In his marginalia and subsequent publications, Viviani portrayed his teacher less as the hallowed initiator of a new physicomathematical tradition and more as the inheritor of an ancient lineage of mathematician-commentators on Euclid. Chapter 3 highlights the ways Mersenne relied on and broke with established modes of reading, adapting traditional textual practices associated with sixteenth-century natural philosophy to Galileo's book.

Chapter 4, "An Annotated Book of Many Uses," extends this analysis by considering the annotating strategies of two other readers unknown or overlooked in the Galileo historiography: the English virtuosi Seth Ward (1617–1689) and Sir Christopher Wren. These readers relied on a variety of strategies to understand and appropriate Galileo's text for specific and disparate goals, some related to traditional natural philosophy and others more akin to developing mathematical and experimental programs. Their practices underline the instabilities inherent in Galileo's text, as readers applied to it multiple approaches, from established Scholastic and humanist practices to new mathematical techniques.

Chapter 5, "The University of Pisa and a Dialogue between Old and New," and chapter 6, "Jesuit Bookish Practices Applied to the *Two New Sciences*," move to more localized and controlled environments of institutionalized philosophy to consider questions not only of scholarly practices but also of pedagogy and the dissemination of received interpretations of Galileo's book. Whereas the earlier chapters focus primarily on extant marginalia, examining how readers interacted directly with Galileo's printed text, these final chapters rely on a variety of sources to uncover indirect evidence of reading. I argue that while the rhetoric of Pisan professors often emphasized the divisions between innovators and traditionalists in their faculty, their reading methods suggest that these lines were not so easily drawn in practice. Jesuit readers, like their Pisan counterparts, relied on the topical reading strategies of Scholastic and humanist learning to read the *Two New Sciences* and to integrate it into their courses on natural philosophy.

Many of Galileo's readers felt the pressure of censorship. Few could forget that they were reading the work of an individual condemned by the Catholic Church, and Italian readers especially undoubtedly weighed their interpretations and writings about Galileo against restrictions imposed by the Roman Church, local authorities, and in some cases brethren of their own religious order. Marginalia may suggest themselves as a privileged window into a reader's direct response to Galileo's text, one more spontaneous and less inhibited by the demands of censorship. Yet scholars in the seventeenth century often regarded their annotations as intellectual products developed with time and care and not casual jottings reflecting immediate reaction to the text. As Viviani's case indicates in chapter 2, moreover, concerns about Galileo's reputation—and perhaps the reader's own—may have shaped how one annotated the *Two New Sciences*.

Issues of censorship and self-censorship are more pronounced in chapters 5 and 6, which consider the reception and reading of the *Two New Sciences* in institutionalized settings. Faced with the facts of censorship and rich archival sources documenting the Jesuit internal censorship system, a common approach underlying recent scholarship in this area has been to probe the limits of these restrictions and to attempt to examine the interplay between an individual Jesuit's desire to embrace novelty and the restrictions imposed by censorship.[101] A similar impulse underlies both scholarship that seeks to identify the "true" affinities of individual Jesuits and scholarship that examines the formal and informal censoring practices that shaped the Society's teaching and intellectual output.[102] This question of the limits of Jesuit censorship and the relationship between scholarly predilection and official restrictions is an important one that has yielded much rich scholarship about the internal workings of the order and the relationship between subgroups within it. However, it is not a question central to the analysis in the forthcoming pages. This is because, I would argue, an inquiry into censorship and self-censorship implicitly denigrates the actual intellectual response of the scholars in question; it assumes that if the restrictions had been removed, such individuals would have produced "worthier" scholarship, that is, scholarship untainted by the restrictions of religious authorities or more conservative members of their order.

These chapters instead show the varied ways period readers experienced the *Two New Sciences*, not as a text of the period's New Science but as a printed book by a famous, controversial author whose interpretation required their active evaluation. The meaning ascribed to the *Two New Sciences* was not fixed by the words on the printed page or even by the inclination, or disinclination, of the reader toward novelty or Galileo. Instead it depended on a variety of factors, from the individual reader's training to the locale and circumstances in which the text was encountered.

Much as Galileo may have wished them to, readers did not come to the *Two New Sciences*, neatly categorize it as "new," and proceed to read it using new practices and for new goals. Instead they approached it with the methods in which they were trained, and if it occurred to them and if it seemed reasonable, they adapted these methods as necessary. It was not the case that Galileo proclaimed that the correct way to do science was to look to nature and not texts and that subsequently his enlightened readers did just that. No gestalt or paradigm shift created a community of "Galileans" who immediately had recourse to a separate set of scholarly tools, standards, and objectives. Instead, readers cobbled together the methods, content, and goals they found useful. These scholars formed alliances, engaged in priority disputes, and formed personal rivalries, to be sure, but on close examination of these readers' engagement with Galileo one is hard-pressed to imagine them donning imaginary hats labeled "Galilean" or "Aristotelian" and rushing off into battle, as Galileo often implied in his writing.[103] For Galileo, the intellectual community was engaged in such a battle, with his opponents roused against him because his discoveries contradicted those physical propositions commonly accepted in philosophical schools.[104] From the perspective of Galileo's readers, however, the landscape was much more fluid. In the words of the Pisan professor Pascasio Giannetti (1661–1742), who largely abandoned Aristotle in his teaching in favor of the approach of Galileo and others, "Philosophers' opinions, various and in disagreement, [have] made this science [of physics] very difficult and even confusing."[105]

Historians currently use the term *New Science* to refer to a variety of enterprises, from Galileo's science of motion to the ideas of Paracelsus. What unifies the New Science is its opposition to the "old," usually an Aristotelianism associated with conservative institutions like the universities and the

Catholic Church. Thus, the New Science has been contrasted with, for example, "Jesuit science," the "learning of the schools," and "Catholic Physics."[106]

Galileo's readers, in contrast, acknowledged generational shifts more than divided groups. They used a variety of terms to describe people doing new things. The Pisan professor Claude Bérigard (1578–1663) labeled Galileo a *neotericus* (a modern, "new" person), *iunior* (a younger person), and *recentior* (a newer, fresher, more recent person). In their teaching texts, Jesuit professors usually referred to Galileo by name and occasionally identified him as belonging to a larger set of philosophers they classified as *recentiores*.[107] When the Pisan professor of natural philosophy Mauro Mancini (who taught philosophy from 1692 to 1703) offered a Cartesian-infused approach to the questions traditional to Aristotelian physics, he titled his course "Aristotle restored, a new philosophy from old philosophies."[108]

These terms range in meaning from "innovator" to "restorer." Even the label *novator*—a word often used to describe individuals like Galileo, though not by the readers considered in the coming chapters—can be translated in a variety of ways. It derives from the Latin *novare*, which means to make new, renovate, renew, refresh, or change, so that a *novator* can be one who renews and restores, as well as one who innovates. Translated in this way, it is not only Galileo who is a *iunior, recentior, neotericus,* or even *novator* but all of his readers too, for they were also recent and modern writers who aimed to renew and restore their period's natural philosophy. This book charts the methods by which such *novatores* made sense of the *Two New Sciences* and sought to integrate it into their own explanations of the natural world.

CHAPTER ONE

An Anonymous Annotator, Baliani, and the "Ideal" Reader

Because writings intended for public consumption always have an audience in mind, a reader's private notes offer a more uncensored window into his or her reaction to a printed text. This chapter and the next three rely on such privileged sources—in particular, heavily annotated copies of early editions of the *Two New Sciences*—to examine the responses of readers to Galileo's book. These readers' marginalia allow us to see them in action, revealing the parts of the text they chose to annotate, the methods they used to read and process the passages, and the questions, sources, and concerns they brought with them to the *Two New Sciences*. These chapters necessarily draw on and enrich the growing body of scholarship related to early modern reading and note-taking practices, and I use this literature to contextualize the methods of Galileo's readers. Through their focus on a text in the mixed-mathematical tradition whose author forcefully proclaimed his methodological novelty, they also extend the current focus of this area of research, one that has concentrated on descriptive genres, often with regard to authors and readers who were decidedly commonplace. This discussion of reading methods is not intended as an end in and of itself, however, but rather as a means of interrogating larger questions of reception and the transformation of scientific ideas in the early modern period.

This chapter uses an annotated copy held today at the National Library in Florence (BNCF) as a starting point for reconstructing the scholarly practices of Galileo's readers. BNCF manuscript Galileo 80 (BNCF Gal. 80) contains a plethora of annotations, both textual and diagrammatic, written in a combination of Latin and Italian. Nearly 170 pages of the 314-page book are annotated. The marginalia are spread roughly equally over the four days of the dialogue, with slightly more annotations in Days 1 and 2 than in Days 3 and 4. The appendix on centers of gravity is not annotated at all.[1] At times the marginalia look as if they were composed in different hands, suggesting that more than one person contributed to them. The marginalia are limited to the actual pages of the *Two New Sciences*. None of the flyleaves, the covers, or the front matter contains notes. Nor, as far as the historical record shows, did this reader insert additional note slips into the copy.

While the annotator of this volume cannot be determined with certainty, I consider the possibility that the annotator was the Genovese patrician Giovanni Battista Baliani, a correspondent of Galileo's who also wrote on questions of local motion.[2] Baliani was an Italian of roughly the same genera-tion as Galileo who shared many of Galileo's intellectual interests and whose opinion Galileo actively sought. Galileo sent Baliani copies of many of his pub-lications, including his 1612 *On Floating Bodies* and his 1613 *Letters on Sun-spots*. Baliani visited Galileo in Florence in 1615 and discussed problems of motion and mechanics, and the two were in correspondence from 1614 to the last years of Galileo's life.[3] Thus, if Baliani did annotate the book, his margi-nalia offer an insider's perspective on how a contemporary close to Galileo read and reacted to it.

The identity of the annotator, however, assumes secondary importance to larger questions about what these annotations reveal about the scholarly practices of this reader. The marginalia in this copy correspond in many ways to what is known and accepted about the reception of the *Two New Sciences* by period readers. Analysis of the copy thus provides the opportunity to ex-plore in detail the standard story of the reception of the *Two New Sciences* and to describe the typical or "ideal" reader expected from the historiogra-phy. After establishing characteristics of this "ideal" reader, I turn again to the marginalia to highlight aspects of the reader's reaction that the scholar-ship does not anticipate. One central feature of this reader's approach is the way he drew on multiple reading practices and norms to read Galileo's text. At times he expressed a concern with the correspondence between Galileo's mathematics and physical phenomena reminiscent of the aims of reform-ing philosophers and mathematicians like Galileo himself. In other pas-sages he read more traditionally though still eclectically, drawing on the established practices of multiple disciplinary traditions. The eclectic meth-ods this reader applied to Galileo's text set the agenda for subsequent chap-ters, which seek to enrich the established narrative of the reception of the *Two New Sciences* by placing it within the context of early modern scholarly practices.

A Text of Mathematical Natural Philosophy

Galileo's decision to apply mathematics to the question of motion, a topic treated qualitatively by Aristotle and central to his natural philosophy, is one reason the *Two New Sciences* has been described not just as a text of mixed mathematics but rather as one of mathematical natural philosophy. In Day 3, Galileo defined uniformly accelerated motion as one that "adds on to itself from rest equal momenta of swiftness in equal times." Galileo also insisted that his quantitative results corresponded to physical reality, arguing that his definition was in agreement with *naturalia experimenta* (physical or natural things experimented or tried).[4] The mixed-mathematical disciplines, including mechanics and astronomy, had since antiquity been devoted to creating mathematical descriptions of natural phenomena. These quantitative accounts were supposed to correspond to what was observed in nature, but they were not always understood as descriptions of how the physical world worked. Instead, the discipline of natural philosophy was charged with answering these questions of physical reality. By purporting to offer mathematical descriptions that corresponded to phenomena as they take place in nature and treating the subject of local motion, a topic traditionally considered the purview of natural philosophers, Galileo positioned the *Two New Sciences* in the emerging category of mathematical natural philosophy.

One key feature of our reader's approach to the *Two New Sciences* is his pressing concern with the physical validity of Galileo's claims, especially regarding the science of motion. This focus on the intersection of quantitative description and physical reality pinpoints what historians have recognized as innovative and important about the *Two New Sciences*. Unlike the modern historian, however, this annotator was not convinced by Galileo's claims to have described physical reality through his quantitative rules. In the margins alongside Galileo's first definition of accelerated motion, our reader immediately noted his doubts regarding Galileo's claims. The reader wrote that while Galileo's earlier definition of equable motion matched what is found in nature, that pertaining to accelerated motion did not. Even more troubling was the fact that Galileo only offered a probable demonstration of its validity. A more prudent manner of exposition, he noted in the margin, would have been to present a sound proof of his definition, one that relied on experience or demonstrative reasoning rather than probabilistic argument. Only such an approach would bring Galileo's readers to believe him unquestionably.[5] Here the reader referred to a common distinction in the period between reasoning based on certain premises, hence demonstratively certain, and conclusions that were "probable," or worthy of approbation, because of the probable status of their experiential premises.[6]

As he proceeded, this reader continued to question whether the definition actually resembled physical reality. At the point where Galileo's interlocutor Sagredo interrupts the reading of the Academician's treatise to note that while any definition can be supposed by the author, he himself has doubts about whether the proposed definition conforms to that found in nature, the anonymous reader underlined Sagredo's words and indicated his own agreement in the margin, noting that Sagredo's doubts matched his own. For the annotator, the issue was "whether that definition is adapted to that accelerated motion with which heavy bodies naturally descend, which is what the Author presupposes."[7]

The reader's perception that Galileo's arguments lacked demonstrative certainty is another consistent theme in his marginalia. At one point in Day 3, Sagredo describes his understanding of what happens when a heavy body falls. Following the rules laid out by the Academician, Sagredo reasons, a body will have obtained eight degrees of speed in eight pulsebeats, four at the fourth beat, and so on. For Sagredo, a paradox arises, for there is no degree of speed so small that the movable will not have possessed it at some time, and yet even at the beginning of its motion the movable appears to move with great speed. Sagredo's concern is with the apparent discrepancy between the observed swiftness of the initial movement and the Academician's claim that motion is initially infinitely slow. Our reader read this paradox as further evidence that the quantitative relationship proposed by Galileo might not match what happens in nature. He noted in the margin, "It is probable but not certain, because it is possible that in the sixth [beat] it would be three, or five [degrees of speed], and similarly in other [beats]. Therefore, I don't know when such a proposition is admitted as a supposition, whether it is a science, or whether it needs to be demonstrated."[8] The reader here returned to the same point that bothered him initially. He agreed that it was probable that in the sixth pulsebeat the movable would have acquired six degrees of speed, but he was not convinced that this quantitative relationship was certain. The initial proposition of the quantitative relationship, according to him, needed to be demonstrated to be admitted as true and to substantiate Galileo's science of motion as a true science.

This anonymous reader summed up his frustration and doubts regarding Galileo's treatment of naturally accelerated motion at the end of Day 3. Day 3 concludes with a conversation among Galileo's interlocutors, who affirm the novelty of the Academician's findings on motion. They marvel at the number of propositions the Academician was able to develop from one single postulate and note their astonishment that ancient thinkers, including Archimedes, Apollonius, and Euclid, as well as many philosophers, had never developed these results. Our reader was not similarly impressed, as he made clear in an extended annotation. He noted first that the claim that all of the results thus presented were deduced from only one very simple principle was not entirely true, for "this treatise depends on those [many] principles and suppositions or petitions as can be seen in [his] marginal notes." What is more, he complained, this single principle that the interlocutors praise "is none other than the definition of uniformly accelerated motion," coupled with the assumption that it agrees with natural motion, an agreement that Galileo never proved. The reader went on to compare Galileo to those authors who dedicate themselves to squaring the circle—a mathematical task that was proved in the nineteenth century to be impossible—and who, in doing so, deduce "many very beautiful conclusions." For such beautiful conclusions, the reader admitted, Galileo and these mathematicians "are worthy of much praise," but this praise did not negate the fact that Galileo never proved his principal point, "that natural bodies of natural motion have and are provided with the proportions that are asserted here." According to him, Galileo had merely demonstrated that such bodies would have these proportions "if in fact they are moved with motion that really goes accelerating uniformly," which of course the reader doubted was true.[9] For this reader, Galileo misrepresented his accomplishments and never proved his main contention.

Similar doubts regarding the relationship between physical phenomena and Galileo's mathematical results plagued the reader as he moved to Day 4. Galileo predicated his analysis on the supposition that a projectile could be analyzed as two compound motions, one naturally accelerated motion in the vertical direction and one equable, or nonaccelerated, motion in the horizontal, and that these would not interfere with each other. Alongside Galileo's exposition, the reader wrote that his "demonstration proceeds with the supposition that the two motions are maintained in the same tenor, as if each one is responsible for itself and provides no impediment to the other." Yet, the reader remarked, "I very much doubt whether that supposition is true; in fact I believe exactly the opposite." He went on to argue that the cause for his doubts stemmed from the fact that if the supposition of independent motion were true, "the projectile's velocity and the force of its blow would grow nevertheless even more." Citing his own observations of projectiles shot with crossbows and of objects falling perpendicularly, he argued that the differences in the force of the blow delivered by projectiles was too great to be the result merely of the air's resistance.[10] In a later passage, the reader returned to this point and emphasized his frustration. "It bothers me," he wrote in the margin, "that it seems possible to suspect that these two motions provide an impediment to each other and that they can alter each other's movement." If such doubts were valid, he noted, the horizontal and vertical motions Galileo assigned to the projectiles would not obtain.[11]

That our reader would be concerned with the physical validity of Galileo's arguments may seem obvious. At the most basic level, wouldn't we expect a reader interrogating an author's claims to have described nature to do so by considering whether he had achieved his goal? Modern readers would answer in the affirmative. Recent scholarship, however, has indicated that early modern readers may not have been so quick to adopt this critical gaze. Sixteenth-century natural philosophy, according to this interpretation, was a bookish enterprise in which readers read books to gather a range of opinions about natural phenomena, not to pronounce on the veracity of textual claims.[12] In the same period, mixed-mathematical reading, at least in the case of astronomy, also often proceeded by focusing on mathematical validity and setting questions of physical reality aside.[13]

In the context of this literature on early modern scholarly methods, our reader's response thus marks him as somewhat unusual. It appears that he did not continue to inhabit the sixteenth-century world of textual scholarship, in which the goal was to amass multiple opinions on a given subject. He wanted to find the right answer, and he was not convinced that Galileo had done so. By interrogating Galileo's assumptions in the margins of his text, this anonymous reader followed Galileo's own advice, constantly questioning the correspondence between words and nature. When criticizing opponents, such as the Jesuit Orazio Grassi, Galileo emphasized that writing about the natural world required investigation into and observation of nature; nature, not writing on a printed page, served to validate men's conclusions. Galileo urged his contemporaries to look beyond the written word and verify textual claims through empirical observation. The interlocutors in the *Two New Sciences* too model this type of approach. Again and again, Salviati leads

Sagredo and Simplicio not to accept what the Academician has written but to consider personal experience and reported experiments to validate what is inscribed in the treatise.[14]

These annotations in Days 3 and 4 thus give us some idea about the characteristics of our reader. First, he was committed to describing physical reality with quantitative rules. Second, he was knowledgeable in the study of local motion and likely had a vested interest in the subject. Moreover, he disagreed with many of Galileo's conclusions and specifically doubted whether Galileo's quantitative descriptions of naturally accelerated and projectile motion corresponded to motion as it took place in the world around him.

Baliani and the "Ideal" Reader

Our reader's criticisms of Galileo's science of motion resemble the science of motion described in another seventeenth-century text, Baliani's *De Motu Naturali Gravium Solidorum*, first published in 1638, with a second, expanded edition in 1646.[15] Comprising only forty-three pages, this first edition did not integrate mathematical descriptions of nature with accounts of experiments as had Galileo in his *Two New Sciences*. Baliani gestured to various observations he had made of the behavior of falling bodies while acting as prefect of Savona in 1611, but he organized the body of the work as a mathematical treatise.[16] Baliani's observations regarding the actual motion of heavy bodies are contained in the definitions, suppositions, and postulates that open the treatise, but he did not justify these assumptions with explicit reference to specific experiments.

In the first edition, of 1638, Baliani's descriptions of local motion closely resembled Galileo's own. He claimed, for example, that accelerated motion follows the odd-number rule, which states that the spaces traversed by a movable undergoing naturally accelerated motion in equal times increase as the odd numbers do, so that if the movable traverses a distance of one unit in the first second of time, it will traverse a distance of three units in the second second of time, five units in the third second, and so forth. Baliani also affirmed that the length of the pendulum's string is proportional to the square of its period.[17] In the 1646 second edition, Baliani left the original *De Motu* largely intact but expanded the work by including five additional books treating the subjects of impetus and the motion of liquids. The collection of additional theorems that he added to the extra days indicates that by 1646 Baliani had departed substantially from Galileo's conclusions. For example, though his statement of the odd-number rule was unchanged, his discussions of impetus suggest that he endorsed an alternative rule of fall formulated by Honoré Fabri, a Jesuit natural philosopher who engaged with the latest currents in philosophical speculation in his writings.[18] Another key feature of Baliani's revised text is his rejection of Galileo's claim that projectiles follow a parabolic path.[19] Both of these aspects of Baliani's later analysis correspond closely to the types of criticisms leveled at Galileo's *Two New Sciences* in the annotated copy.

There is, moreover, an additional hint that further confirms the possibility that the annotator may have been Baliani. Alongside Proposition 2 in Day 3 the reader criticized Galileo for not proving his definition of naturally accelerated motion but only assuming it, noting that "if it [uniformly accelerated motion] were given, this characteristic would be sought, and this nowhere in fact was proved to be given, which I proved in my treatise *de Motu naturali.*"[20] This annotation appears to reference the reader's treatise *De Motu Naturali.* With no other information given, it is difficult to identify with certainty the text cited. However, the parallel between the title cited in the annotation and the titles of Baliani's two editions, *De Motu Naturali Gravium Solidorum* (1638) and *De Motu Naturali Gravium et Liquodorum* (1646), is unmistakable. Given the correspondence between Baliani's writings on motion and the opinions expressed in the annotations, this reference to a treatise *De Motu* strongly suggests that Baliani was indeed the annotator of BNCF Gal. 80. The variability of the handwriting both in the 1638 copy and in Baliani's surviving letters, however, makes this identity uncertain, especially given the widespread practice of employing amanuenses in the period.[21] For this reason, I refer to this reader as "pseudo-Baliani" in the ensuing discussion. This designation is intended to convey the possibility, but not the certainty, that the copy may have been annotated by Baliani. This tentative identification also serves the purpose of anchoring the ensuing discussion of the "ideal" reader to a concrete historical figure.

Pseudo-Baliani's annotations and the historical Baliani corroborate and reinforce much of the accepted narrative of the reception of the *Two New Sciences* in the historiography. This narrative has created an "ideal" seventeenth-century reader with four central characteristics. First, like pseudo-Baliani and the historical Baliani, well-known readers of the *Two New Sciences* have been concerned with the veracity of Galileo's claims. In his synthesis of recent literature on early modern mechanics, Domenico Bertoloni Meli outlined the response of readers across Europe to Galileo's work on mechanics contained in the *Dialogue* and *Two New Sciences*.[22] The responses of these readers varied from Pierre Gassendi's experiments involving dropping stones from the masts of moving ships to the Jesuit Pierre Le Cazre's formulation of an alternative rule of fall. According to historians' accounts, the single question unifying the responses of these readers was, what is the correct rule of fall? Implicitly, these readers were intent on verifying whether Galileo's rule of fall was the right one. To decide what was "right," they adopted a variety of criteria. Some, like Marin Mersenne and Giovanni Battista Riccioli, sought experimental confirmation of the rules. Others, including the Jesuits Niccolò Cabeo and Fabri, concentrated more on the theoretical underpinnings of Galileo's findings, questioning, for example, Galileo's assumptions about the composition of the continuum. For readers like Galileo's students Torricelli and Vincenzo Viviani, who agreed that Galileo's rules were correct, the question then became, how can we strengthen their mathematical foundations? Despite their different approaches, all these readers, according to current portrayals, agreed that there was one "right" rule of fall, that it was important to determine what this rule was, and that there existed criteria—whether experimentation, mathematical validity, or philosophical rigor—that could be employed to determine whether Galileo's rule was the correct one.

Second, and very much related to the first characteristic, experiment or observation of the physical world was a key element for these readers in verifying Galileo's claims. Again, Bertoloni Meli's synthesis serves as a useful guide for demonstrating the dominant role of experiment in accounts of the reception of Galileo's *Two New Sciences*. More than half his chapter on the reception of Galileo's work describes experiments carried out to test Galileo's theories. These experiments included the historical Baliani's work with collaborators to drop heavy bodies from the masts of moving ships in the Genoa harbor to see where they landed and Gassendi's experiments to do the same off the coast of Marseilles. A number of readers, including Baliani, Mersenne, and Riccioli, attempted to measure the time and distances of bodies in free fall or along inclined planes, while others carried out experiments with artillery to validate Galileo's claims about projectile motion. Other experiments inspired by the reading of the *Two New Sciences* included attempts to measure the frequency of vibrating strings and the periods of pendulums, tests of the speed of water exiting a pierced cistern, studies of the rarefaction and condensation of air, and trials to determine the resistance of wires to breaking.[23]

Third, as the above account already hints, the "ideal" reader of the current historiography exhibited an overwhelming interest in Galileo's science of motion, contained in Days 3 and 4 of his *Two New Sciences*, compared with his science of the resistance of materials. In the words of Bertoloni Meli, echoing the assumptions of current scholarship, Galileo's science of motion "attracted the lion's share of the interest."[24] Again, the historical Baliani's response to Galileo's work reinforces this image of the "ideal" reader. Baliani's publications, the two editions of his *De Motu Naturali*, were contributions to the science of motion that paralleled and built on Galileo's own. Baliani's letters to Galileo detailing attempts to verify experimentally Galileo's findings also focused on Galileo's descriptions of falling bodies and the lengths and times of their falls.[25]

Finally, a fourth characteristic of the standard narrative of the reception of the *Two New Sciences* has been the close association in the minds of readers between the 1638 work and the earlier, prohibited *Dialogue*. This strand of the narrative in effect implies that the "ideal" reader read the *Two New Sciences* hand in hand with the *Dialogue* and saw Galileo's arguments about motion and the resistance of materials as integral to his support of the Copernican heliocentric system.

There are sound reasons for making this assumption. Galileo himself linked the two publications structurally. In the *Dialogue*, three interlocutors, named Salviati, Sagredo, and Simplicio, meet over the course of four days to debate the merits of the Ptolemaic and Copernican systems. Salviati, identified by Galileo with Filippo Salviati (1582–1614), to whom he also dedicated his *Letters on Sunspots*, presents arguments on behalf of the heliocentric system. Simplicio, whose name calls to mind both the Italian word *sempliciotto* (simpleton) and Simplicius of Cilicia (ca. 490–ca. 560), the late antique commentator of Aristotle, defends those of Ptolemy. Sagredo serves as a discerning intermediary and stands as a tribute to Galileo's friend and former pupil Gianfrancesco Sagredo (1571–1620).[26]

The same interlocutors grace the pages of the *Two New Sciences*. In this later publication, they play no such predefined roles, though Salviati, in general, acts as the pupil of and spokesperson for the unrepresented character of the Academician, and Simplicio is often called upon to speak for Aristotle and his commentators. The continuation of the dialogue between these three characters was recognized by period readers. For example, Galileo's contact in Bohemia, Giovanni Pieroni, warned Galileo that it might not be in his best interest to use the same characters that he had employed in the *Dialogue*.[27]

Galileo too offered a preview of some of his findings on motion in the *Dialogue* and promised his readers a forthcoming publication that would treat the topic in greater detail. In Day 1 of the *Dialogue*, for example, Galileo's interlocutors discussed his finding that heavy bodies acquire the same velocity when moved along inclined planes of the same height.[28] In Day 2, further speculations were presented in response to Sagredo's query to Salviati about whether he had ever "toyed with the investigation of these ratios of acceleration in the motion of falling bodies." Salviati responded with a direct allusion to many of Galileo's findings on local motion: "I have not needed to think them out, because our common friend the Academician already showed me a treatise of his on motion in which this is worked out along with many other questions. But it would be too great a digression if we were to wish to interrupt our present discussion for this." Sagredo refrained from asking more, but he did express his wish that Salviati would examine these propositions "in some special session."[29] Salviati acknowledged Sagredo's desire at the end of the *Dialogue* when he promised to return for "one or two further sessions," in which he would communicate "elements of our Academician's new science of natural and constrained local motions."[30] These "further speculations" came in Days 3 and 4 of the *Two New Sciences*, where they are presented in the guise of a mathematical treatise composed in Latin, written by this same Academician. They are read aloud by Salviati to his interlocutors. Historians have argued that these references in the *Dialogue* were not just self-promotion but expressed the deeper connections between Galileo's support of Copernicanism and the assumptions underlying his science of motion.[31]

Galileo's contemporaries are also thought to have recognized these conceptual links. One of the most important readers in this regard was Pierre Gassendi, who in 1642 published his *De Motu Impresso a Motore Traslato*, in which he made explicit the connections between Galileo's assumptions about motion and the Earth's motion. Gassendi's treatise in turn prompted a widespread debate, carried out in print and in manuscript correspondence, on the validity of Galileo's rules of fall. The vociferous nature of the debate led Paolo Galluzzi to name this episode the "second Galileo affair." Unlike the first "Galileo affair," which focused on Galileo's condemnation, this one was in response to Galileo's laws of motion, which were "submitted to a series of severe censures between 1642 and 1648."[32] The controversy enveloped scholars across Europe, Catholics and non-Catholics alike, and involved such seventeenth-century luminaries as Mersenne and his circle, including

Descartes, Pierre Fermat, and Christiaan Huygens. Galileo's former students and correspondents in Italy, including Baliani, contributed to the debate, as did a number of Jesuit scholars, including Le Cazre, Fabri, and Riccioli.

Pseudo-Baliani and historical Baliani thus throw into relief important assumptions underlying scholarship on the reception of the *Two New Sciences*. The received narrative paints a picture of an "ideal" reader assumed in the historiography. Such a reader was interested in the relationship between textual claims and physical reality; attempted to verify Galileo's claims, often experimentally; focused his attention on Galileo's science of motion; and read the *Two New Sciences* in connection with the 1632 *Dialogue on the Two World Systems*. This figure of the "ideal" reader reinforces the interpretation of the *Two New Sciences* disseminated more widely in the historiography, one that portrays the text as groundbreaking in terms of content and methodology. It does so by assuming implicitly that Galileo's proclamations of innovation made the work tangibly novel to his readers. According to the portrait of the "ideal" reader, Galileo's readership for the most part had broken substantially with the textual, descriptive scholarship associated with the Aristotelian tradition; they read and interpreted his text with the modern goals of verification and the tools of experiment and mathematics. If the text was not modern and novel on its own, its links with Galileo's support of Copernicanism heightened readers' sensitivity to its claims, making them realize the cosmological implications of the work's less glamorous subject matter.

Verification and Experiment

While the historical Baliani has been a key figure in standard accounts of the reception of the *Two New Sciences* and fits the mold of the "ideal" reader well, pseudo-Baliani's marginalia reveal a more tenuous correspondence. It is true that pseudo-Baliani did exhibit great interest in the correspondence between Galileo's mathematics and physical phenomena. Other aspects of his marginalia, however, reveal that he read the *Two New Sciences* in surprising ways not described in the literature.

For one, his annotations suggest a more complicated relationship between text and experiment than we might assume based on the accepted narrative. The only reference in his marginalia to a specific real-world experience is an oblique citation to his own experience observing the relative speeds of objects falling perpendicularly from fifty feet.[33] In other instances, in contrast, pseudo-Baliani sought to verify Galileo's descriptions of experiments not by carrying out the procedures described by Galileo but by comparing textual accounts. Alongside one passage in Day 1, for instance, he noted that the ratio of 400 to 1, reported by Galileo as the ratio of the heaviness of water to the heaviness of air, was too high and that a ratio of 50 to 1 was more appropriate. His argument was based not on his own experimental trials but on the application of calculational principles and statements found in Galileo's text. Earlier in Day 1, Galileo had compared the falls of various objects, including lead, ebony, and oak, in air and in water by calculating how many degrees of speed each object lost to the surrounding medium as a result of its heaviness. Alongside Galileo's computations of the falls of these materials, the reader carried out his own reckoning of the relative speeds of lead and wax. He then turned to a separate passage many pages earlier, where Galileo had declared that wax weighed nearly the same as water. Based on Galileo's assertion there and on his own calculations carried out in the margins, the reader concluded that in a fall of fifty feet, the lead would reach the ground, while the wax would remain about a foot behind.[34] Using these numbers and the principles Galileo employed to calculate the comparative velocities of materials in different media, the reader claimed that Galileo had overestimated the ratio of the heaviness of water to the heaviness of air.

This focus on textual comparison, as opposed to experimental verification, is surprising given the emphasis on the latter in accounts of the reception of Galileo's *Two New Sciences*. This absence is even more noteworthy given that there was an established tradition of using the margins of printed books to record observations. European medieval readers commonly annotated astronomical texts and tables with their own astronomical observations. John of Murs, who taught at the University of Paris in the fourteenth century, for example, recorded such observations as marginal notes in a 226-folio collection of astronomical texts and tables. His objective was to compare his own observations of eclipses, conjunctions, solstices, and equinoxes with calculations based on the Alfonsine Tables.[35] In the sixteenth century, Paul Wittich annotated Copernicus's *De Revolutionibus* with various observations, including conjunctions of Mercury and Venus with different stars.[36] Since Baliani did report carrying out experiments, it is possible that our annotator— whether or not he was the historical Baliani—did carry out experiments but opted to record his findings in separate notes. We know, for example, that early modern physicians compiled observations from their medical practice and added to them notes derived from their correspondence and reading, often organizing these patient histories into compilations for future reference.[37] Practicing alchemists and readers of medical recipes engaged in a similar process, creating separate notebooks in which they recorded textual excerpts and the results of their own experimental trials.[38] Yet, the absence of references to experiment in pseudo-Baliani's marginalia at the very least hints that when he read the *Two New Sciences*, he did so without immediate attention to experimental verification.

There is also evidence that pseudo-Baliani read for more than just verification of Galileo's claims. His annotations, for example, demonstrate a significant interest in Galileo's mathematics for their own sake, not just for the physical results they purported to describe. His typical approach was to indicate in the margins the relationship between the steps of Galileo's proofs and the earlier propositions and theorems on which they depended. Consider, for example, his annotations to Proposition 17, Problem 4, of Day 3, in which Galileo asks his readers to find the point E on the diagram (fig. 1.1) such that a body's fall through AB (starting from rest) would take the same amount of time as the body's fall along BE (starting from the velocity acquired by the body after it fell through AB). Pseudo-Baliani did not focus his annotations on Galileo's conclusion, which stated that the required distance BE could be determined by constructing BF equal to AB and setting the point D such that BD was to DF as DF was to DE. Instead, he read to follow along with Galileo's proof, inserting letters in the text of Galileo's proof that were keyed to marginal notes. In these notes, he indicated which of the propositions proved earlier in the text allowed Galileo to draw the conclusions that he did. Thus, for example, the reader noted with the letter *a* Galileo's statement that "if AB is assumed to be the time through AB, the time through DB will be DB." The letter *a* is keyed to the marginal annotation that reads, "by his third proposition."[39] Earlier in Day 3, Galileo had presented his third proposition, which stated that the time it takes a movable to fall from rest along an inclined plane and the time it takes that movable to fall along a vertical are proportional to the lengths of the plane and the vertical, respectively.[40] What interested pseudo-Baliani was Galileo's mathematical logic, not his final conclusions.

In other cases, pseudo-Baliani followed along by rewriting the steps of the proof in the margin. As an example, consider how he responded to a problem presented at the beginning of Day 2 (fig. 1.2). Here, Galileo considered the resistance of bodies to breaking and argued that such problems were reducible to the law of the lever. Readers were asked to determine how much of the total weight of the body was sustained on the horizontal plane and how much by the force at the end of the lever. Galileo proved this assertion by

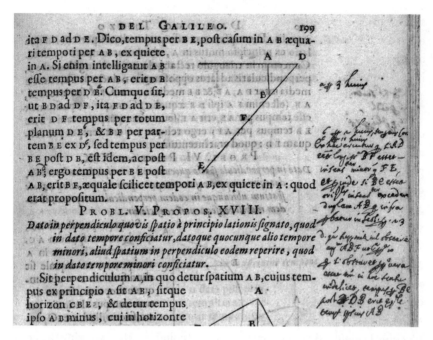

Fig. 1.1. Pseudo-Baliani's annotations to Proposition 17, Problem 4, of Day 3. *BNCF Gal. 80, 199 (detail). Courtesy of the Ministero dei beni e delle attività culturali e del turismo / Biblioteca Nazionale Centrale di Firenze.*

manipulating a number of proportional relationships based on the law of the lever. Pseudo-Baliani began his annotation with a summary of the steps Galileo used to solve the problem: "The moment of stone A to the moment at G has the proportion compounded from GN to NC and FB to BO, it is as FB to BO, in this way NC [is] to X, and the moment of the rock to G [is] as GN to X, because if the power B to C is as FO to OB, and by compounding the moment A to the power G is as FB to BO, [which is as] NC to X. The power C to G is as GN to NC, therefore the moment A to G is as GN to X."[41] Effectively, our reader reproduced in words Galileo's proof, mimicking the prose style of Galileo's presentation by summarizing it in the margins. Again, pseudo-Baliani read to follow along with Galileo's proof. He was more interested in the mechanics of Galileo's demonstration than in the conclusions proven by it.

In the descriptive portions of the *Two New Sciences*, pseudo-Baliani often read to translate Galileo's claims into the terms of Aristotelian natural philosophy. In Day 1, Galileo argues that the continuum must be composed of infinitely many indivisibles, because if a line could be divided without end, it would have to be divisible into infinitely many parts. Pseudo-Baliani wrote in the margin that the line " might not be [divisible] with an end but it could be [divided] into a syncategorematic infinity, as they say in the schools, but it never could be [divided] into an actual infinity."[42] Here pseudo-Baliani expressed his disagreement with Galileo, who had argued for the possibility of an actual infinity. He related Galileo's text to the university context by employing the term *syncategorematic infinity* and noting that it was an expression used "in the schools." Following medieval precedent, early modern commentators on Aristotle differentiated between the "categorematic" and the "syncategorematic" infinite. These terms derived from Scholastic logic and corresponded loosely to the distinction between the actual and the potential infinite. Taken categorematically, the infinite referred to an actually infinite number. In contrast, the syncategorematic infinite corresponded to

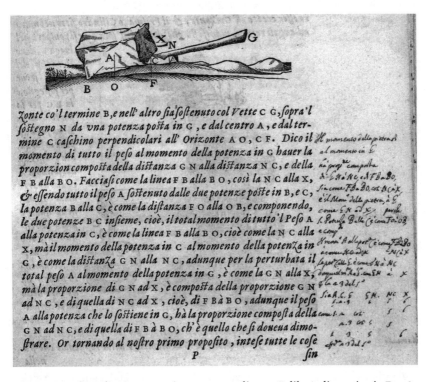

Fig. 1.2. Pseudo-Baliani's annotation corresponding to Galileo's discussion in Day 2 showing that the problem of resistance of bodies to breaking is reducible to the law of the lever. *BNCF Gal. 80, 113 (detail). Courtesy of the Ministero dei beni e delle attività culturali e del turismo / Biblioteca Nazionale Centrale di Firenze.*

a potentially infinite number.[43] Pseudo-Baliani believed that Galileo's claim for an actual (categorematic) infinite was incorrect but that a syncategorematic infinite was possible.

As the dialogue proceeded, he continued to relate Galileo's discussion to the Scholastic tradition. Pseudo-Baliani remarked repeatedly that whatever Galileo claimed about the infinite required proof and could only apply to the syncategorematic or potential infinite. Thus, Galileo's claims that the continuum was composed of infinite indivisibles, that there was no infinite number other than unity, and that a line that was continually divided eventually would comprise infinitely many parts all needed additional proof and were not substantiated by the text.[44] Similarly, this reader objected to Galileo's claim that apart from infinite or finite, there existed the category of answering to every designated number. Pseudo-Baliani noted that this designation was not precisely correct but would better be understood as conforming to the established category of the syncategorematic infinite.[45]

Our reader also demonstrated an adherence to the aims of Aristotelian philosophy by showing a marked interest in the causes of natural phenomena. In an annotation to a passage in Day 1 claiming that water could be raised no higher than eighteen braccia using a suction pump, he suggested that it was not the case that this behavior was the result of nature's abhorrence of the void, as Galileo argued. Rather, he noted, it was possible that the rise in the water was the result of the pump's rarefying the water and that eighteen braccia was the maximum to which the water could be rarefied.[46] The coherence of the filaments of a rope, which Galileo attributed to their mutual compression, was explained for pseudo-Baliani by the antagonism between their parts. He provided a mechanical explanation for this antagonism, attributing it to the ruggedness or roughness of the parts of the filaments, which impeded their motion.[47] In these examples, the reader took the descriptions reported in the text as true and only questioned Galileo's explanation of their underlying cause. This attitude toward textual accounts of natural phenomena resembles that ascribed to natural philosophers working in the Aristotelian tradition, who held that demonstrations could only be based on generalized, common experiences and regarded the development of causal explanations as their ultimate goal.[48]

Galileo's account of water on cabbage leaves provoked a similar reaction. According to Galileo, water on cabbage leaves tended to bead up, rather than to dissolve and flow over the leaves, because of a great dissension between air

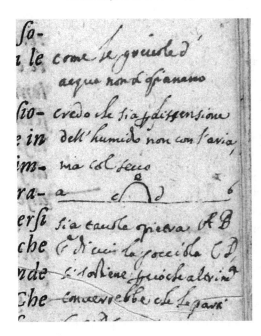

Fig. 1.3. Pseudo-Baliani's annotation corresponding to Galileo's explanation of the behavior of water on cabbage leaves. *BNCF Gal. 80, 71 (detail). Courtesy of the Ministero dei beni e delle attività culturali e del turismo / Biblioteca Nazionale Centrale di Firenze.*

and water. In his annotation, pseudo-Baliani first noted that he did not agree with Galileo's causal explanation. Rather than because of a dissension between air and water, he stated, the water formed droplets as a result of the dissension between moisture and dryness.[49] He augmented this first explanation, which attributed the behavior of the droplets to the contrariness of the Aristotelian qualities of moistness and dryness, with a second account that appealed to mechanical properties. His description was aided by a diagram that he included in the margin of Galileo's text (fig. 1.3). According to this second explanation, "Let the table or rock be AB, on which the drop CD supports itself for this reason that otherwise it would be suitable that the parts C and D flow toward A and B, and therefore on top of [something] dry, their contrary. In other words, [it is] because the table is rough, over which the water has difficulty flowing, and it is less difficult [for the water] to support itself in this way than to overcome the difficulty of flowing over roughness."[50] In his annotation, pseudo-Baliani equated the two explanations, the one based on the contrariness of dryness and moistness and the mechanical one of the roughness of the table. He focused on the causal explanation Galileo attributed to the phenomenon and sought to provide an alternative one in which mechanical features provided an underlying basis for Aristotelian principles of contrary qualities.

Despite his embrace of a mechanical explanation, pseudo-Baliani returned to the style of traditional, pedagogical discourse by interrupting his annotation to provide the response of an imaginary interlocutor. At the bottom of the page, he wrote, "You will say that the water E ought to flow over the surface EC or ED." "I respond," he noted, "that the water D, like a heavy body, does not tend to C or D but to the center of the world."[51] This style of objection and response calls to mind the similar organization of humanist and Scholastic pedagogical dialogues. These dialogues were based on the *quaestio* format used in university teaching, in which a series of objections were proposed to the explanations proffered and then resolved.[52]

Whereas the ideal reader portrayed in the historiography read Galileo to verify his findings with experiment, pseudo-Baliani read for a variety of purposes and rarely referenced his own experiences. His reaction to the science of motion in Days 3 and 4, in which he interrogated Galileo's quantitative description with respect to motion as it takes place naturally, is an example of only one aspect of his overall response to Galileo's text. In other parts of the book, he read for the steps of Galileo's mathematical proofs, not their conclusions, and he was interested in Galileo's causal explanations and in translating Galileo's opinions into the standard philosophical language of the schools. Experiment figured little, if at all, in his annotations, and textual comparisons often served as the means of verifying Galileo's reported findings.

Galileo's Science of Motion and Copernicanism

The annotations in pseudo-Baliani's copy also cast doubt on the assumption that Galileo's readers were most interested in his science of motion and read the *Two New Sciences* in light of his earlier *Dialogue*. Whereas pseudo-Baliani annotated more than 60 percent of the pages in Days 1 and Day 2, only about 50 percent of the pages in Days 3 and 4 contain marginalia. These numbers alone suggest that pseudo-Baliani was at least as interested in Galileo's science of materials as he was in his science of motion.

Pseudo-Baliani's annotations also contain no mention of the *Dialogue*. This is surprising given the prevailing practice of using one's marginalia to draw attention to parallel passages in related works.[53] Because cross-referencing related works was such an established part of early modern note-taking practices, it is natural to assume that if readers saw the *Two New*

Sciences as related to the Dialogue, they would have noted such a relationship in the margins of the two books. Moreover, my partial census of extant copies of the Two New Sciences has unearthed no such cross references in copies of the first or second edition of the text. Furthermore, partial censuses of the Dialogue reveal, similarly, no mention of the Two New Sciences in that publication.[54] The only such cross reference between the Dialogue and the Two New Sciences I have encountered in marginalia is in the annotations Mersenne inscribed in his copy of his Harmonie universelle. Here, Mersenne cited the Two New Sciences alongside the printed passages in which he discussed Galileo's Dialogue.[55]

Given this lacuna in the reading record, it is beneficial to reconsider briefly the evidence on which the assumption that Galileo's two publications were read in tandem is based. One of the main reasons that scholars assume that the reception of Galileo's Dialogue and the reception of his Two New Sciences are entangled is the scholarship on the second Galileo affair, the debate on Galileo's science of motion initiated by the publication of Gassendi's 1642 De Motu.[56] Examination of the sources that contributed to this episode reveals, as expected, a scholarly community that discussed both the Dialogue and the Two New Sciences. There is indication, however, that individuals in this community did not always read and cite both of Galileo's works in tandem in their contributions to the larger discussion.

Gassendi's De Motu, for example, links Galileo's science of motion and Copernicanism but appears to be based largely on Galileo's Dialogue. Throughout the text, Gassendi refers generally to "Galileo," without specifying the book whose conclusions he describes.[57] The majority of the examples he cites clearly are drawn from Day 2 of Galileo's Dialogue. These include Galileo's explanation of why pendulum bobs eventually come to rest, his calculation of the time it would take a stone to fall from the Moon's or the Sun's orbit to the center of the Earth, and his theory of the tides.[58] Such content is fitting, given Gassendi's purpose in considering the relationship between Galileo's science of motion and Copernicanism. There is only one passage in which Gassendi appears to have articulated a claim not found specifically in the Dialogue. Here, however, it is unclear whether Gassendi is indebted to the Two New Sciences.[59]

Gassendi's reliance on the Dialogue contrasts with one of the earliest reactions to his De Motu, a letter composed by Pierre Le Cazre (1589–1664), rector of the Jesuit college at Dijon, and sent to Gassendi in November 1642.[60] Le Cazre responded directly to Gassendi's letter, citing specific corollaries and examples from it. He recognized the key, explosive element of Gassendi's argument, namely, the way Gassendi had linked a discussion of Galileo's science of motion with a defense of Copernicanism. However, unlike Gassendi, Le Cazre discussed Galileo's science of motion by turning directly to the *Two New Sciences*. The first three paragraphs of Le Cazre's letter address Gassendi's discussion of Galileo's findings on naturally accelerated and projectile motion. Here Le Cazre cites the *Two New Sciences* explicitly, in one instance drawing his reader's attention to page 164 of Galileo's "third dialogue on accelerated motion," as well as to "that entire third Dialogue."[61] Le Cazre, who opposed the Copernican hypothesis, also found several of Galileo's claims regarding motion problematic, including Galileo's odd-number rule. Although he found Galileo's findings on projectile motion more palatable, Le Cazre also criticized several propositions regarding the time traveled and the force possessed by objects following a parabolic trajectory.[62]

Despite these censures, it is not clear how Le Cazre judged the relevance of his disagreement over the quantitative rule of fall to the larger philosophical and cosmological concerns he felt Gassendi raised in his letter. After exploring the science of motion in his first three paragraphs, Le Cazre focused the rest of the letter on philosophical and cosmological concerns. In Le Cazre's words, "But since it is not important to your purpose how, in the end, the acceleration of heavy bodies proceeds, either arithmetically or geometrically, I do not devote more attention any longer [to it] here."[63] While LeCazre certainly disagreed with Galileo's science of local motion, then, he dismissed the question of its veracity as irrelevant to a host of other issues, including Gassendi's atomism, his hypothetical defense of the Copernican system, and his adherence to Galileo's theory of the tides, all issues he saw as more pressing. In these later sections of his letter Le Cazre turned to the *Dialogue*, criticizing Gassendi's endorsement of Galileo's theory of the tides, but here he did not cite the 1632 text directly, referring instead to Gassendi's previous missive. In fact, in his entire letter Le Cazre avoided direct citation of any text other than Gassendi's letter and Galileo's *Two New Sciences*.

Riccioli's 1650–51 *Almagestum Novum* offers further indication that individuals who debated Galileo's science of motion largely drew on only one of Galileo's publications. The *Almagestum Novum* was intended as a comprehensive guide to mid-seventeenth-century astronomical knowledge. Riccioli began work on the text in 1640, not only relying on the Jesuit libraries in Parma and Bologna to cull material but also assembling a group of students and colleagues to carry out astronomical observations and experimental trials involving freely falling objects, projectiles, and pendulums. The massive, two-tome text was complete by 1646, but difficulties with Jesuit censors delayed its publication until 1650.[64] Riccioli's treatment of astronomical topics was comprehensive; he made a point of including specific citations of every past and contemporary author who had discussed the subject at hand. Yet in the sections in which he addressed the topics of local motion and the Copernican system, Riccioli never cited the *Two New Sciences*. His discussion of Galileo's science of motion remained firmly tied to the treatment in the *Dialogue*, even though Riccioli cited earlier publications by Galileo related more to mechanics than to astronomy, namely, Galileo's 1612 *On Floating Bodies*. Riccioli did not even cite the *Two New Sciences* when he treated topics discussed in greater depth in the 1638 text than in the *Dialogue*, including the behavior of pendulums and falls down inclined planes.[65]

These patterns suggest that the *Dialogue* and the *Two New Sciences* were not always linked in the minds of readers, indicating some limits to the discourse involving both of Galileo's books. Participants who brought both books into dialogue often drew on only one text in their own work. Le Cazre's response to Gassendi indicates that period readers may have judged debates over the mathematical ratio of falling objects to be incidental to the larger philosophical and cosmological concerns that drove their scholarship. It is possible that while the second Galileo affair was prompted by concern about the cosmological implications of Galileo's science of motion, individual historical actors may not have read the *Dialogue* and the *Two New Sciences* in tandem.

These conclusions suggest the need to expand current assumptions about Galileo's readership and to modify our narrative about the reception of his *Two New Sciences*. The following chapters explore these themes in greater depth and also investigate the models for the types of reading observed. Other annotators, even those closely associated with the New Science and the promotion of Galileo's work, similarly read the *Two New Sciences* with many of these and even other goals. Moreover, a closer examination of their reading practices will reveal that many of their methods derived from common sixteenth-century approaches. By exploring the reactions of multiple readers with different attitudes toward Galileo, associated with a variety of institutions, based in locales across Europe, and of various religious affiliations, I will show that this eclectic approach to Galileo's *Two New Sciences* was not an aberration but the norm. When faced with Galileo's claims to novelty in terms of both methodology and content, readers responded by drawing on a variety of reading practices and assessing Galileo's text in light of a range of philosophical opinions. Galileo's mechanics was disseminated not because a small group of committed Galileans embraced wholeheartedly his methods and conclusions but because a wider audience read and made sense of his work in a more piecemeal fashion, appropriating it into less coherent and more eclectic scholarly projects.

CHAPTER TWO

Editing, Commenting, and Learning
Math from Galileo

Vincenzo Viviani became personally acquainted with Galileo when he went to Arcetri as a pupil and amanuensis following Galileo's condemnation. Arriving in October 1639, Viviani was not present during the exchanges between Galileo and his contacts in Venice and northern Europe that led to the publication of the *Two New Sciences*. However, through his association with the Grand Duke of Tuscany, Ferdinand II (1610–1670), who recommended him to Galileo, Viviani was personally acquainted with the circle of Galileo's Italian admirers who eagerly awaited the 1638 book. While Galileo had offered a preview of his findings on local motion in his *Dialogue* and promised more details in a future publication, he also circulated some of his results in manuscript and correspondence. Galileo's student Bonaventura Cavalieri, for example, relied on Galileo's quantitative rules of free fall, which Galileo had shared with him earlier, in his own *Lo specchio ustorio*, published in 1632.[1] Readers also had access to Galileo's treatise *Le mecchaniche*, which remained unpublished by Galileo in his lifetime but circulated in Italy, England, and France.[2] When Viviani arrived in Arcetri, he thus entered the home of a revered philosopher and mathematician whose contributions to astronomy and mechanics would have been well known to him.

Often remembered as "Galileo's last student," Viviani is a principal actor in the standard account of the reception of the *Two New Sciences*, included for his devotion to his late teacher and for the volume of textual material he preserved and produced. In this chapter, then, we move from Baliani, a scholar who was a contemporary of Galileo's, to another reader, also close to the master in terms of intellectual development and cultural context but of a younger generation. Viviani is often considered in tandem with another of Galileo's students, Evangelista Torricelli. Torricelli and Viviani served successively in Galileo's former position as mathematician to the Grand Duke of Tuscany. They also shared an interest in solidifying the foundations of

Galileo's work on mechanics, especially his science of motion. The *Two New Sciences* was central to their efforts. Torricelli revisited the *Two New Sciences* in his 1644 *Opera Geometrica*. In the section "De motu," Torricelli set out to offer a new and firmer basis for Galileo's science of motion, to expand on his teacher's results, and to provide a more extensive treatment of parabolas.[3] Also concerned with the mathematical underpinnings of Galileo's science of motion, Viviani assisted the aged Galileo, recording a proof for the axiom underlying Day 3 that his teacher dictated to him. This proof was added to Day 3 in the 1655–56 reprint of Galileo's collected works, which Viviani helped see into print.[4] After Galileo's death, both Viviani and Torricelli continued to work with Galileo's unpublished manuscripts. They developed some of these writings into two additional "days" to augment the 1638 publication. The Fifth Day was published in 1674 by Viviani as *Quinto libro degli Elementi d'Euclide, ovvero Scienza universale delle proporzioni* (The fifth book of Euclid's *Elements*, or A universal science of proportions).[5] A Sixth Day, containing Galileo's work on the force of percussion, was elaborated by Viviani on the basis of manuscripts held by Galileo's son, Vincenzo Galilei.[6]

One attribute that distinguishes Viviani from Torricelli is the former's dedication to preserving Galileo's findings for posterity and to promoting his reputation as a pious natural philosopher and mathematician. Viviani petitioned tirelessly for the construction of a commemorative tomb for Galileo at the church of Sante Croce in Florence. Galileo's condemnation proved a constant point of struggle for Viviani, blocking his efforts to erect the tomb. It also shaped contemporaries' accounts of Galileo in ways that Viviani found distasteful. In an effort to control this discourse, he responded by keeping abreast of contemporaries' discussions of and writings about his teacher.[7] Viviani's own biography of Galileo, which remained unpublished in Viviani's lifetime, stressed not the condemnation but the approaches to scholarship that Galileo himself emphasized in his writings, namely, his break with the textual tradition and his reliance on experiment and quantification.[8]

This chapter focuses on an artifact that survived thanks to Viviani's efforts to record and preserve traces of Galileo's life: a copy of the 1638 edition filled with Viviani's marginalia.[9] Viviani's identity as the annotator is indicated by the initials *VV* on the verso page of the copy's title page.[10] I use the annotated copy to uncover the methods Vivian employed as he read the *Two New Sciences*. I inquire how Viviani moved from Galileo's text to the responses to it that he shared with contemporaries in letters and in print. A reader typically included in studies of the reception of Galileo's *Two New*

Sciences, Viviani falls squarely within the camp of the "ideal" reader explored in chapter 1, someone who promoted Galileo and his program and worked to extend and strengthen Galileo's study of mechanics and motion. This chapter looks beneath the surface of these intellectual projects to reveal that Viviani was more closely tied to older methods of scholarship than we might assume. Despite Viviani's apparent embrace of Galileo's novelty, Viviani read the *Two New Sciences* largely in isolation from the *Dialogue* and approached the book in ways that resemble the methods of sixteenth-century readers, especially of mixed mathematics.

Webbing the Text in Layers of Commentary

Viviani's annotations reveal a close attention to Galileo's text achieved by repeated readings and the use of active note-taking strategies, two methods often ascribed to sixteenth-century readers. Several annotations reveal that Viviani returned to the *Two New Sciences* even more than ten years after it was published. For example, in one note slip inserted into his printed copy, Viviani recorded that the slip pertains "to page 114 of the Leiden edition of 1638 and to page 86 of the Bologna edition of 1656."[11] This annotation confirms that Viviani was still taking notes on the *Two New Sciences* after the latter date. This type of reading and rereading has traditionally been linked to more descriptive texts, for instance, in the realms of traditional natural philosophy, theology, or history. Plato's first readers in fifteenth-century Florence often read his writings meditatively in the model of Saint Benedict's *sacra lectio,* in which a contemplative rereading of the text served as a means of approaching the divine.[12] A reader who had more in common with Viviani might have been Gabriel Harvey, who returned again and again to his copy of Livy, seeking material with which to provide political counsel to his patrons at the Elizabethan court.[13] The reading, commenting, and teaching of Aristotle in the early modern university also followed a similar model, in which professors and teachers repeatedly revisited the writings of Aristotle and his commentators.

Viviani's note-taking extended beyond the margins of the *Two New Sciences.* His surviving copy contains numerous paper slips that were later pasted in it. That these slips were once loose sheets is evident from the careful labels Viviani included on most of them, indicating the author, work, and other identifying information. Viviani began one note slip that suggested an addition to the dialogue with the label "In the *Discorsi e dimostrazione matematiche di Galileo Galilei* from the printing of Leiden of 1638."[14] Such labels suggest that Viviani employed note slips more generally in his reading of other texts. If Viviani had been concerned only with the *Two New Sciences* or even only with works by Galileo, he would not have needed to indicate the author, title, and publication date on his note slips.

Certain features of Viviani's annotations suggest a multistep process of taking notes, indicating that the resulting marginalia are not records of Viviani's immediate, unfiltered reaction to Galileo's text but a carefully conceived intellectual product in their own right. His marginal annotations are written in a combination of pen and pencil. In several instances, Viviani wrote the same annotation twice, once in pencil and once in ink.[15] It is likely that Viviani first wrote out his initial impressions in pencil and then returned to the text, perhaps in some cases after consulting with Galileo, and rewrote his annotations in pen. Viviani also employed this practice with diagrams, drawing new figures first in pencil and then in ink.[16] In certain instances, Viviani might have annotated in pencil when his conclusions were tentative.

The presence of additional notes on the *Two New Sciences* by Viviani also confirms that Viviani relied on an elaborate system of note-taking, of which the annotations in the 1638 copy are only one part.[17] These pages include a sheet on which Viviani composed short notes on various sections of Galileo's *Two New Sciences*. He organized them as a list that indicates the page number in the 1638 edition and then his own observations, including summaries of the topics Galileo discussed, the contributions of Galileo's students to the findings, and possible additions to the text. Some of these remarks correspond to aspects of the text that Viviani also pointed out in his marginal annotations, adding further weight to the hypothesis that the annotations in the printed copy were the product of an earlier set of notes.[18]

Elaborate note-taking schemes were a common feature of early modern reading, one emphasized in schools. Sixteenth-century Parisian school texts are often abundantly annotated in the margins and on interleaved pages. Like Viviani's marginalia, these notes are neat and were likely copied over after being initially recorded in the classroom.[19] The *Constitutions* of the Jesuit colleges, first drafted in 1548–50, advised students to read the relevant text before coming to class. During lectures, they were to keep a notebook or sheets of loose paper in which they recorded what was said by the professor; more advanced students would return to these notes and recopy them later in private study. Following lectures, groups of two to four students would engage in repetition exercises based on the notes.[20]

These practices were applied in the mixed-mathematical sciences too. The sixteenth-century Urbino mathematician Guidobaldo del Monte (1545–1607), for example, annotated the sections of his copy of Giovanni Battista Benedetti's (1530–1590) 1585 *Diversarum Speculationum Mathematicarum et Physicarum Liber*, which dealt with mechanics. To understand passages he found especially problematic, he took additional notes and constructed diagrams in a separate notebook that he titled *Meditatiunculae*.[21] Over the course of the sixteenth century, readers of Sacrobosco's *Sphere* devoted increasing attention to the text's mathematical operations, filling its margins with calculations described in the printed text and accompanying tables.[22] Readers of Copernicus's *De Revolutionibus* returned to the printed book again and again, working through mathematical and diagrammatical calculations, indicating relevant astronomical observations, and noting important points in the margins.[23] Viviani's marginalia suggest that he approached the *Two New Sciences* in much the same way as these earlier readers had approached their reading. He saw Galileo's book as one to which these bookish methods of careful note-taking and rereading were suitable.

Viviani also found value in the annotations of other scholars. On note slips and the back flyleaves of his copy, he included several long passages that he indicated had been copied from two other annotated copies of the 1638 edition. One was that of Father Clemente Settimi (b. 1612), of the Scuole Pie, Viviani's first teacher of mathematics, who had introduced him to Galileo and who often visited Galileo at Arcetri. In his copy, Viviani transcribed four of Clemente's corrections and additions pertaining to Day 1, indicating that he had found these annotations in the margins and on loose sheets in Clemente's copy of the 1638 edition. Viviani obtained the other notes from a copy of the 1638 edition owned by Galileo's grandson Cosimo di Vincenzio Galilei and annotated both by Cosimo di Vincenzio and by Galileo's amanuensis Pier Ferri. According to Viviani, all of these additions expressed the desires of Galileo, who had been unable to record them himself because he was blind in the last years of his life.[24]

The practice of transcribing the annotations of others was common in the period. In the sixteenth century, Paul Wittich (ca. 1546–1586) copied the annotations of Erasmus Reinhold (1511–1553) in multiple exemplars of Copernicus's *De Revolutionibus*.[25] This practice continued in certain circles through the eighteenth century. In the 1750s, for instance, a philologist in Amsterdam named Pieter Fontein (1707–1788) transcribed the annotations Isaac Casaubon (1559–1614) had carefully inscribed in a 1541 edition of Theophrastus's

Opera Omnia into his own recently purchased copy of the same edition.[26] What is of interest in Viviani's case is that he was a reader attuned to and a participant in the creation of new scientific knowledge in the later seventeenth century and one who consciously celebrated the novelty of Galileo's methods, yet he approached his teacher's text using many of the techniques common to sixteenth-century readers. One might imagine, for example, that someone interested in the latest scientific developments would be reading new sources by 1656, but Viviani's annotations indicate that he found Galileo's *Two New Sciences* worthy of study long after its initial publication. The continuities between Viviani's approach to the *Two New Sciences* and established textual practices also stand in contrast to Galileo's explicit rejection of traditional approaches. The fact that notes by Viviani on other books survive, including the 1533 edition of Jordanus of Nemore's *De Ponderibus*, moreover, indicates that Viviani's note-taking on the *Two New Sciences* was a central part of his scholarly practice, one that complemented and facilitated his own intellectual work.[27]

Reading Along with a Master

The contents of Viviani's marginalia reveal that he focused his note-taking on the same types of information as did sixteenth-century readers. Viviani, for example, provided topical and summary notes in the margins. Alongside Galileo's speculation that the fluidity of heated metals was a result of subtle fire particles' penetrating void spaces between elements of metal, Viviani wrote "things melted" in pencil in the margin.[28] When Galileo discussed the possibility of the continuum being composed of infinitely many indivisibles, Viviani similarly scrawled "atoms."[29] Alongside a passage explaining that different ratios of the oscillations made by two strings produce different chords, Viviani copied the names of the chords and their ratios, noting, "Diminished fifth or semidiapente," "diapason, double [ratio], octave," "Diapente, Fifth, Sesquialter [ratio]," and "Diatesseron."[30] These definitions seem to have been a point of interest for many readers. An extant copy of the 1638 edition in Milan is clean apart from an extended note by one reader on the inside back cover in which all these definitions are noted.[31]

Viviani also shared with sixteenth-century readers of both textual and mathematical genres an interest in the relationship between the *Two New Sciences* and other texts. In the course of one proof in Day 1, Galileo referred readers to Luca Valerio's 1604 *De Centro Gravitatis Solidorum*. Viviani indicated in the margins that the same demonstration could be found in the first book of Archimedes's *On the Sphere and the Cylinder* and more generally in his *On Conoids and Spheroids*.[32] In Day 3, Viviani noted that Galileo's proof of Proposition 29 could be cut short by adding a reference to Euclid, for "this is proved adequately by itself . . . by the twenty-fifth proposition from the fifth book of the *Elements*."[33] Occasionally Viviani noted how parts of the *Two New Sciences* related to one another and to other writings by Galileo. Alongside Galileo's justification in Day 4 for considering the projectiles shot by firepower differently than those shot by slings, bows, or catapults, Viviani noted in pencil, "Here is repeated what was said on page 93."[34] In Day 3, when Galileo makes a passing comment that Proposition 6 could also be proved another way, "according to mechanics," Viviani wrote that Galileo himself had shown the same result in an earlier treatise.[35]

Such a use for marginalia, not to record personal responses to the text but to facilitate retrieval and retention of interesting passages, was a common feature of early modern reading practices. The most common type of marginal note in Latin books in the period was that flagging passages of interest. Such marginalia could be nonverbal, indicated by underlining or marginal marks, or by keywords highlighting the topics treated, the examples mentioned, or the authorities cited in the passage in question.[36] Cross-referencing passages within and between books was also common, both in pedagogical and scholarly practice. Bartholomew Dodington (ca. 1535–1595), who was the Regius professor of Greek at the University of Cambridge from 1562 to 1585, for example, littered the margins of his teaching text on rhetoric with a wealth of cross references, copying out pertinent passages from Cicero, Aristotle, Sturm, and Plutarch.[37] These practices were employed by readers of illustrated books and mixed-mathematical texts as well. Thomas Lorkyn (1528–1591), a professor of physics at Cambridge, annotated his copy of Vesalius's 1555 *De Fabrica* with an abundance of notes and created his own internal cross-referencing scheme to connect Vesalius's text and images.[38] Del Monte at times annotated his copies of Benedetti and Jordanus of Nemore by indicating the authors' sources and summarizing their arguments.[39] Gingerich and Westman have shown how Copernicus's readers, including Erasmus Reinhold, were "remarkably aware of Copernicus's sources . . . even when Copernicus did not credit them explicitly."[40]

Other features of Viviani's reading suggest his application of established practices oriented specifically to mathematical reading. Already this specialization was visible in the examples described above in which Viviani cited sources that changed or shortened Galileo's proofs. In these cases, Viviani

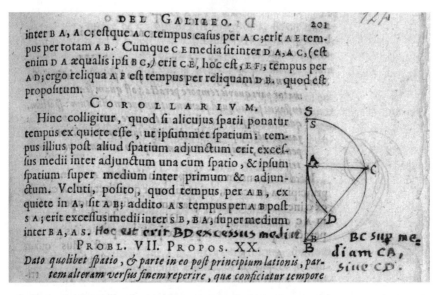

Fig. 2.1. Viviani adds to Galileo's diagram and text to offer a proof of the corollary. *BNCF Gal. 79, 201 (detail). Courtesy of the Ministero dei beni e delle attività culturali e del turismo / Biblioteca Nazionale Centrale di Firenze.*

was not naming Galileo's sources per se but instead identifying alternative argumentative strategies that did not merely augment the text, as the citation of similar passages would have done, but rather could transform it.

In other instances Viviani, like pseudo-Baliani, was interested in the mechanics of Galileo's mathematical proofs and often developed alternative solutions to Galileo's demonstrations. In Day 4, for instance, Viviani developed a different method of calculating the sublimity of a parabola, or the height above the parabola from which a heavy object must fall before being turned horizontally at its peak in order to describe the given parabola.[41] Viviani also devised additional corollaries to Galileo's propositions.[42] Sometimes he added to the text and diagrams to extend Galileo's results or to complete proofs of corollaries that Galileo had merely stated rather than proved explicitly. In a corollary following the nineteenth proposition of Day 3 (fig. 2.1), for example, Viviani proved Galileo's result by following the same steps that Galileo had employed in his solution to an earlier problem in the book. With his textual annotations and additions to Galileo's diagram, Viviani reapplied the steps Galileo had used to solve the previous problem, providing a demonstration of the corollary and simultaneously working through the mathematical logic underlying both the problem and the corollary. Here we can draw a parallel with the way Wittich annotated Copernicus's *De Revolutionibus*. Similarly to how Viviani proposed alternative proofs and corollaries, Wittich explored diagrammatic arrangements beyond those provided by Copernicus's text, including one that may have been the inspiration for Tycho Brahe's geohelio-centric planetary model.[43]

Viviani also followed along with Galileo's demonstrations by employing a combination of textual and diagrammatic annotations in a fashion analogous to that of pseudo-Baliani. At the beginning of Day 3, Viviani noted which line segments in Galileo's figures represented the physical characteristics distance (writing *Sp* for *spatium*), time (*Te* for *tempus*), and velocity (*Ve* for *velocitas*) (fig. 2.2). At other times he keyed conclusions reached by Galileo in the course of his proof to results proved in earlier propositions.[44] Again, there are sixteenth-century precedents for these practices. Del Monte, for example, added labels to diagrams printed in his edition of Jordanus of Nemore and relied on these annotations as he followed along with the accompanying proof.[45] At Oxford Henry Savile (1549–1622) employed a keying system similar to those of pseudo-Baliani and Viviani to note the relationship between the intermediate conclusions reached by the mathematical authors he read and the earlier results on which these conclusions depended.[46]

The attention Viviani paid to Galileo's diagrams was characteristic of period mixed-mathematical reading more broadly. Alexander Marr has shown that Bonaventura Cavalieri's discussion of Archimedes's legendary mirror prompted an exchange of letters between Mutio Oddi (1569–1639) and his friend Piermatteo Giordani. Oddi's initial response was a mathematical proof accompanied by a diagram, though he promised to test the phenomenon as soon as possible.[47] Heinrich Glarean's (1488–1563) students also employed diagrams as they read, working out his rules of musical theory in the margins of his publications. Some even included musical examples, complete with verse and notes, illustrating the different modes (or scales) described in the text.[48] Copernicus's readers too drew their own diagrams in the margins and flyleaves of their copies of *De Revolutionibus*; while these diagrams often replicated those figures found in the printed edition, at other times readers moved beyond Copernicus's text to design new representations.[49] That Viviani was not alone in continuing these practices into the seventeenth century is apparent from surviving copies of Galileo's earlier *Dialogue on the Two World Systems*. In a copy held by the Buffalo Museum of Science, an unknown

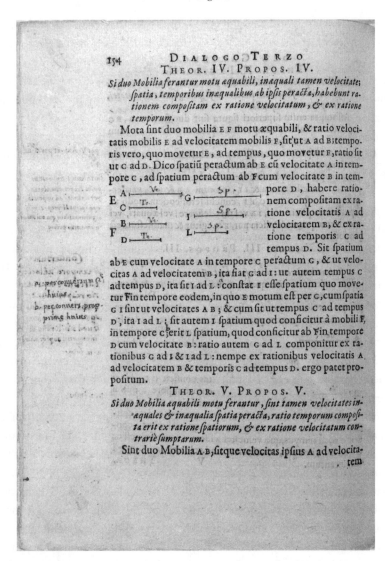

Fig. 2.2. Viviani's annotation of diagram and text to follow along with Galileo's proof. *BNCF Gal. 79, 154. Courtesy of the Ministero dei beni e delle attività culturali e del turismo / Biblioteca Nazionale Centrale di Firenze.*

reader included additional diagrams at the back illustrating the Copernican system, retrograde motion, and the annual motion of the Earth.[50]

All of these features of Viviani's note-taking reveal that he drew on well-established practices when he read the *Two New Sciences*. This conclusion may seem obvious to those familiar with the literature on scholarly methods, a field that has emphasized the longevity of the period's reading and note-taking practices. To scholars of Galileo, however, this point is worth emphasizing because Galileo was so vocal in proclaiming his embrace of novel practices, because he vociferously urged his readers to eschew traditional ones, and because scholarship on Galileo and especially the *Two New Sciences* has tended to emphasize its aspirations to novelty. That Viviani, who took such pains to promote Galileo's reputation and further his scholarly program and who read the *Two New Sciences* alongside his teacher, annotated his book with traditional methods is thus surprising. Viviani's annotations indicate that the bookish methods that have been studied painstakingly by scholars intent on recovering traditional approaches were also applied to writings intentionally portrayed as novel.

The application of these traditional approaches also implies that Viviani's conception of and response to the *Two New Sciences* differed from those offered by modern historians. Viviani differs in important ways from the "ideal" reader of the historiography described in chapter 1. Whereas most historians have focused on Galileo's novel methodologies and science of motion, assuming that Galileo's contemporary readers did too, Viviani read much of the *Two New Sciences* as a book of mathematical exercises. The multiple problems and demonstrations of Days 3 and 4 served, for Viviani, as a way to follow along with Galileo's geometrical proofs and practice them alongside his teacher. Characterized in this way, Viviani's response suggests additional parallels with Copernicus's sixteenth-century readers, especially Wittich, who, Westman and Gingerich argue, saw Copernicus's text "as a corpus of geometrical constructions."[51] For Viviani and Wittich, reading, respectively, Galileo and Copernicus involved working through mathematical constructions, redrawing diagrams, identifying and correcting errors, and going beyond the mathematics in the text to develop new models and proofs. At least in the note-taking record that survives, it did not involve spending much time on the physical implications of the findings.

The style of Viviani's annotations varies according to the days of the *Two New Sciences*. In general, Viviani employed more mathematical annotating techniques in his reading of Days 2 through 4 and approaches common to a variety of sixteenth- and seventeenth-century genres in his reading of Day 1. This pattern of annotation suggests that Viviani viewed the different days of the *Two New Sciences* as belonging to different genres rather than as parts of a coherent whole to which the same type of scholarly methods could be applied. Again, this contrasts with the assumptions of modern scholars, who have spent considerable energy puzzling over the ambiguous relationship of the four days of the *Two New Sciences*.[52] Here again we can note a comparison between Viviani and earlier readers of Copernicus, some of whom, like Reinhold, focused their annotations on the technical portions of the text and left the cosmological sections virtually unglossed.[53] This reliance on established reading practices, the tendency to read the text as a mathematical exercise book, and an inclination to view it as composed of multiple genres suggest that for Viviani the *Two New Sciences* was not Galileo's testament to a new style of mathematical philosophizing. Rather, it was a text that spoke to older modes of scholarship whose value lay in its being the last publication of a revered master.

Reading to Edit

One of the projects to which Viviani dedicated himself was the publication of his teacher's collected works. Along with several professors at Pisa, he helped see into print the first collected *Opere* of Galileo's publications, minus the prohibited *Dialogue*, which was issued in 1655 and 1656.[54] Viviani's marginalia suggest that perhaps as a consequence of this and similar projects, he read the *Two New Sciences* with the aims and eye of the textual editor or corrector, a profession born with the rise of the printing press.[55] One preoccupation Viviani expressed in his annotations was with Galileo's reputation and legacy, a concern that colored his other activities related to collecting Galileo's manuscripts. On the title page of his annotated copy, Viviani inserted the epithet "Noble Florentine" after the line "Linceo."[56] When Salviati cautioned his listeners that his own speculations on the nature of indivisibles and infinites were a product of human caprice, in contrast to the true and sound judgments of theological doctrine, Viviani enthusiastically praised Galileo, noting that this passage presented a "heroic sentiment, and [one] belonging to a philosopher who is more than Christian, because it is Catholic and very holy, and not of a hypocrite."[57] A similar concern with Galileo's legacy is expressed in another annotation Viviani added to Day 3. Alongside Salviati's comment that he had taken down additional observations of the Academician that he did not have time to share with his fellow interlocutors, Viviani noted eagerly that this passage promised further speculations by Galileo.[58]

Correcting and editing the *Two New Sciences* were also a focus of Viviani's marginalia. Many of the notes in Viviani's copy are concerned with the overall structure of the book. Viviani noted, for example, that the text should be thought of as containing three, not two, new sciences. According to Viviani's

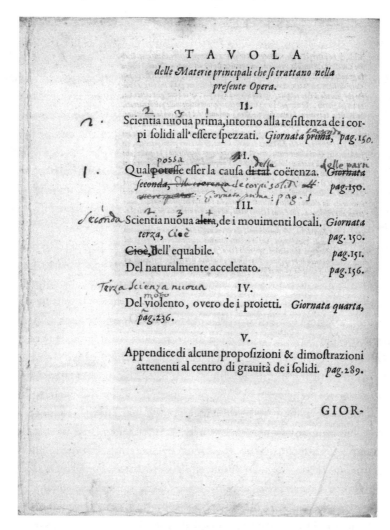

GIOR-

Fig. 2.3. Viviani's annotations to the table of contents. *BNCF Gal. 79, [6v].*
Courtesy of the Ministero dei beni e delle attività culturali e del turismo /
Biblioteca Nazionale Centrale di Firenze.

corrections (fig. 2.3), Day 1 of the *Two New Sciences*—erroneously represented on the title page as Day 2—was not a new science at all but instead concerned the cause of the coherence of the parts of solid bodies.[59] The first new science was presented in Day 2 and treated the resistance of solid bodies. The second new science was that of local motion, both equable and naturally accelerated. And Galileo's treatment of violent motion in Day 4 constituted, for Viviani, a third new science. Viviani carried these corrections through the text, noting, for instance, at the beginning of Day 4, that this was the "third new science."[60]

Viviani also made annotations directly in Galileo's text, aggressively changing word order and phrasing throughout the book. Some of these changes derived from Galileo's own observations. For example, Viviani corrected all the printers' errors noted in the table of printers' errors at the end of the 1638 edition. But he did not stop there. Figure 2.4 shows a typical example of Viviani's vigorously marking and modifying Galileo's prose. On this page Viviani added missing articles (*le* in front of *lunghezze*), altered Galileo's word choice (replacing *quello* with *un pendulo*), and made the text more specific (adding *reciproca* after *la proporzione*).

Other changes Viviani proposed were more substantial. For example, at the end of Day 3 Viviani included an annotation that would have transferred part of the dialogue from Sagredo to Simplicio but then struck out the alteration.[61] In other instances, he suggested adding phrases emphasizing how repeated experiences confirmed Galileo's speculations.[62] Such allusions to experience may have been intended to strengthen the demonstrative nature of Galileo's science of motion. It has been argued that Galileo wrote Days 3 and 4 to present his findings, which were based on mathematical, discrete event-experiments, as conforming to expected ideals of formal, deductive demonstrations, which Aristotle had specified were to be grounded on repeated, generalizable sense experiences. Galileo, according to this reasoning, complied with this Aristotelian ideal by asserting that the results of his experimental trials had been observed to occur "often" and "many times."[63]

Many of Viviani's proposed alterations were intended to formalize and standardize the structure of Galileo's presentation. For example, throughout Days 2 and 3 and the appendix, Viviani identified many of Galileo's unlabeled declarations as propositions, corollaries, and lemmas. In some cases he augmented Galileo's numbering system or inserted his own. For instance, in the 1638 edition, the first eight propositions in Day 2 were numbered, but the final ones were not. Viviani enumerated these remaining propositions in pencil.[64] Aside from an initial lemma, the propositions in the appendix were unlabeled. Viviani went through and categorized them as propositions, corollaries, and lemmas.[65] Here, Viviani's activities as editor and student merged, for he then relied on these labels as he worked through Galileo's mathematics, marking in the margins where Galileo's proofs depended on earlier results. While the structure and numbering system served the more immedi-

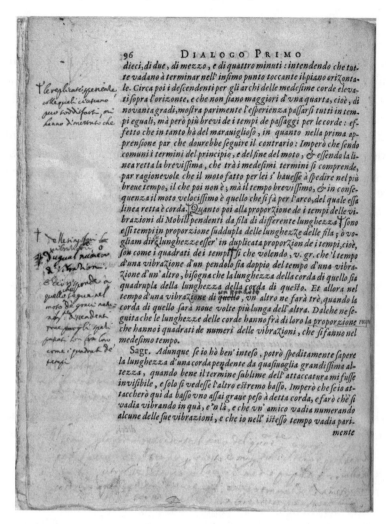

Fig. 2.4. One sample page of Viviani's aggressive editorial interventions in Galileo's text. *BNCF Gal. 79, 96. Courtesy of the Ministero dei beni e delle attività culturali e del turismo / Biblioteca Nazionale Centrale di Firenze.*

ate function of facilitating Viviani's own reading and comprehension of the text, they also may have been intended to do the same for future readers of a new edition.

As a reader and potential editor, Viviani saw the *Two New Sciences* not as an inviolate, fixed text but as one that could be improved through engagement. While it is now well accepted that early modern readers did not identify fixity as a central characteristic of print, it is worth drawing attention to this aspect of Viviani's reading for two reasons.[66] For one, it is a useful reminder that despite Viviani's interest in Galileo's future reputation, he remained firmly entrenched in the world of the early modern scholar. Second, Viviani's continual efforts to improve Galileo's text resemble the approaches of the contemporary English naturalists John Aubrey (1626–1697) and John Evelyn (1620–1706), whose scribal methods encouraged them to view books, information, and descriptions as never complete and always in progress. Though the approaches of Aubrey and Evelyn have been attributed to their adherence to a Baconian program and especially suited to Baconian descriptive genres, the example of Viviani points to the ubiquity of these scribal methods in early modern scholarship, extending even to mathematical works.[67]

Commenting on the *Two New Sciences*

Viviani's scholarly projects extended beyond collections of Galileo's publications and included print editions of Galileo's unpublished writings and commentaries on Galileo's works. The *Two New Sciences* was central to these endeavors, and many of Viviani's publications can be seen to have grown directly or indirectly out of his reading and annotating of that work. One example is found in his 1718 *Trattato delle resistenze*, which was seen into print after Viviani's death by his colleague at Pisa Guido Grandi. Grandi worked from a manuscript of Viviani's that is no longer extant, explicitly reordering Viviani's original material and inserting his own explanatory notes and additions.[68] The longer title of the treatise, *Trattato delle resistenze principiato da Vincenzio Viviani per illustrare l'opere del Galileo* (Treatise on the resistance of materials begun by Vincenzio Viviani to illustrate Galileo's works), reveals Viviani's intention in composing his notes, namely, "to illustrate," or make clear, Galileo's findings.

For Viviani, illustrating Galileo's treatise meant commenting, explaining, and strengthening the presentation in Days 1 and 2 of the *Two New Sciences*. In some instances, Viviani's illustration of Galileo's text involved a condensing of Galileo's dialogue into shorter propositions. In one of these, which Grandi labeled the eighth proposition, Viviani transformed a longer exchange between Galileo's three interlocutors in Day 1 into a concise statement: the resistance of a material to breaking is due both to the material contained within the solid and to nature's fear of the void.[69] He then described an example culled from Day 1 in language that resembled Galileo's own, namely, the behavior of two smooth slabs that resist separation when pulled perpendicularly.[70]

Viviani's reading and note-taking appear to have played an integral part in producing this summary. He inserted a marginal annotation in his copy of the 1638 book clarifying that Galileo's description of the slabs applied to the separation of the superior slab from the inferior.[71] Even more telling, the relevant passages from Day 1 regarding resistance to breaking and the void summarized in this proposition were marked by Viviani in his annotated copy by a series of three quotation marks in pencil. These marks correspond to Galileo's statement regarding the causes of resistance and the details of the example of the marble slabs (fig. 2.5). Marking one's text in this way to indicate passages for extraction was a common technique. August, future Duke of Brunswick and founder of the library at Wolfenbüttel that now bears his name, flagged passages in the books he read, using underlining and manicules. He then copied these passages into separate notebooks.[72] By moving from these marks to the account he offered in the *Trattato*, Viviani applied to Galileo the common technique of summarizing or epitomizing a text, a technique advocated by Renaissance pedagogues and usually discussed in relation to more descriptive genres, especially ancient literary sources.[73]

While Viviani's style of note-taking and exposition is expected in the context of the Scholastic and humanist traditions of summarizing and commenting on texts, the commentary Grandi inserted in the *Trattato* suggests that by the treatise's publication in 1718 such a style of reading was not commonplace, at least not in the genre of experimental philosophy. Following the example of the polished slabs, Grandi told his readers that this section clearly derived from very early drafts of Viviani's manuscript. If it had been composed or revised later, claimed Grandi, it would reflect Torricelli's well-known demonstration of 1644 showing that what Galileo had attributed to the force of the void was nothing but the result of the pressure of the air.[74] Grandi recognized that this section derived from Day 1 of Galileo's *Two New Sciences*, but he expected Viviani's project to be an exposition of current experimental and philosophical opinion, not a summary of or commentary on the *Two New Sciences*. Whether or not Grandi's hypothesis regarding the dates of this proposition is true, Grandi's need to excuse Viviani's text speaks to his conception that reading and writing about Galileo in the first decades of the eighteenth century ought to entail a quest for up-to-date verified propositions about the natural world. The evidence pointing to Viviani's reliance on methods of excerpting and commenting suggests that Viviani likely was engaged in an alternate enterprise.

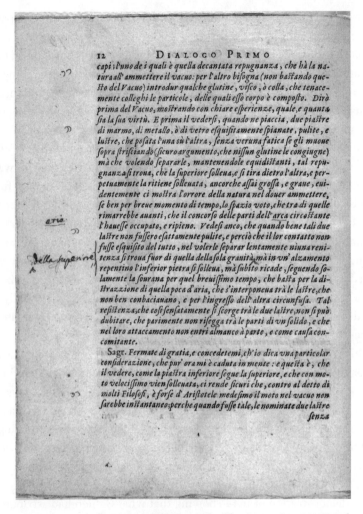

Fig. 2.5. Passage in Day 1 marked by Viviani for extraction. *BNCF Gal. 79, 12. Courtesy of the Ministero dei beni e delle attività culturali e del turismo / Biblioteca Nazionale Centrale di Firenze.*

Making connections through cross-referencing was another technique Viviani employed to illustrate Galileo's text. Near the beginning of his treatise, for example, Viviani explained to the reader how to understand Galileo's notion of the resistance of a solid "taken absolutely." In Day 2 Galileo had explained that his analysis of bodies' resistances to breaking could be understood in two different ways. One could consider objects in the abstract and separated from matter, what Galileo termed "taken absolutely." Alternatively they could be understood as material, as having weight and thickness, or "joined with matter, moment, or compound force."[75] Viviani annotated this section of the *Two New Sciences* extensively, demonstrating his strong interest in it. Alongside most of Day 2's propositions, Viviani added in pencil a marginal note indicating whether the analysis considered the solids absolutely or as possessing weight. Viviani indicated these assumptions by writing "without weight" or "with weight" in the margins.[76]

As in the previous example, Viviani's exposition in the *Trattato* involved a summary of Galileo's text. Viviani explained that the resistance of a solid taken absolutely could be calculated by determining the weight that would break the solid, either in a straight line or transversely. In the first case, the object would be attached above and the weight hung below it, whereas in the second, one end of the object would be attached to a wall and the weight at its other end, pulling perpendicularly to it.[77] Viviani here weaved together two largely separate discussions in the *Two New Sciences*. The first, treated in Day 1, concerned the causes of a body's resistance and considered the resistance of a body to breaking in the vertical direction. In Day 2 Galileo addressed the calculation of a body's resistance to breaking by applying the law of the lever to bodies positioned horizontally and broken transversely by a weight oriented perpendicularly. The two subjects, of course, were linked in Galileo's text; at the beginning of Day 2 Salviati compares the two situations, noting that the vertical breakage of Day 1 is achieved by a much larger weight than is the transverse one.[78] Viviani highlighted these connections by drawing together the two discussions explicitly in his *Trattato*.

Viviani augmented his account with a diagram that was intended to illustrate a method he proposed of testing the absolute resistance of a glass cylinder (fig. 2.6). According to Viviani, in order to determine the weight required to break such a glass cylinder, one should begin with two solid, smooth slabs, designated by AB. Through a hole in AB should be passed a glass rod (CD) whose end is attached to a weight E, which at the point of breakage will be a measure of the glass rod's resistance. Through this device, Viviani wrote, the ratio of the absolute resistance of equal sections of metals could be found.[79]

Though this procedure is not found explicitly in Day 2, the image accompanying it bears a striking resemblance to another device portrayed in Day 1 (fig. 2.7). In this section, Galileo describes a contraption that he claims allows one to determine by how much force nature prohibits a void. It consisted of an empty cylinder of metal or glass (CABD) into which the wooden cylinder

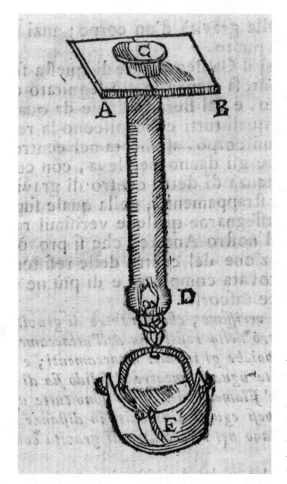

Fig. 2.6. Diagram illustrating a method of testing the absolute resistance of a glass cylinder. *Grandi and Viviani,* Trattato, 5 *(detail). RB 708047, The Huntington Library, San Marino, California.*

EGFH fit. An iron wire with a conical screwhead, which acted as a stopper, was then passed through a hole drilled through the center of the cylinder, and a bucket was attached to it at hook K. To conduct the experiment, explained Salviati, one first filled the cylinder, turned upside down, with water. The cylinder and stopper were inserted and the stopper drawn back. Finally, the apparatus was turned on its head, and a bucket was attached to the hook K. Heavy materials were placed in the bucket until the upper surface of the cylinder (EF) detached from the lower surface of the water. Similar to Viviani's use of the weight E (fig. 2.6), Galileo claimed that the combined weight of the iron rod, the container, and heavy material in the bucket was the force of nature's repugnance to the void.[80]

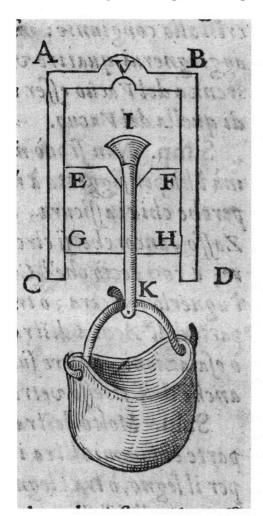

Fig. 2.7. Galileo's device for
measuring the strength of the
void. Two New Sciences *(1638),*
15 (detail). RB 701317, The
Huntington Library,
San Marino, California.

The relationship between the two devices is clear. In both instances, weights are placed in the bucket until a separation is achieved. The total weight in the bucket becomes a measure of the force required to separate or break either the glass rod itself or the stopper from the cylinder. This separation, in turn, indicates the strength of the body's absolute resistance or the strength of nature's fear of the void. The two examples are further linked because Galileo had attributed some of a body's resistance to breaking to nature's abhorrence of the void.

I postulate, then, that Viviani developed this example by reading with an explicit eye to the connections within Galileo's text. As already noted, cross-referencing was a common practice more generally in the period, one observed in the marginalia that survive in texts of both more descriptive and more mathematical genres annotated by sixteenth-century readers.[81] In other passages, Viviani's annotations make clear that he employed cross-referencing as he read. For example, he noted that discussions of the medium's resistance in Day 4 were related to earlier musings on the same topic in Day 1.[82] Although no annotations in Viviani's copy are directly related to the relationship he posited here between Galileo's discussion of absolute resistance and this device in Day 1, Viviani did annotate both of these sections with enthusiasm.[83]

Finally, as he did in his marginalia, Viviani included in the *Trattato* alternative propositions and demonstrations intended to strengthen and augment Galileo's text. Viviani's Proposition 5, for example, relies on a diagram found in the *Two New Sciences* (figs. 2.8 and 2.9). The proposition claims, "For cylinders and prisms equally thick and unequally long, the resistances to breaking transversely are in the reciprocal proportion of their lengths; or, better said, the forces required to break these solids have the reciprocal proportion of the said lengths."[84] Viviani proves this proposition by invoking Galileo's assumption that the beam attached to the wall can be treated as a lever, with arms AB and BC. According to Grandi, Viviani had intended to place this proposition after the first of Galileo's in Day 2 dealing with the resistance of bodies to breakage. It was needed because its contents were assumed in Galileo's fifth proposition, which related prisms' and cylinders' resistances to fracturing to their lengths and thicknesses.[85]

Though it depends on assumptions and diagrams in the *Two New Sciences*, Viviani's proposition is not found in Galileo's text. It is a creation by Viviani, in much the same way as were the corollaries, additional demonstrations, and theorems he inscribed in his copy of the *Two New Sciences*. Again, devising these alternative demonstrations was a common practice among sixteenth-century readers, including del Monte and Copernicus's annotators.

Galileo as the Inheritor of an Ancient Textual Tradition

That Viviani's scholarly work was directed toward preserving and commenting on the writings of textual authorities, in whose number he counted Galileo, is seen even more clearly in Viviani's 1674 *Quinto libro degli Elementi d'Euclide*. This publication, whose subtitle translates as "A universal science of proportions explained by Galileo's doctrine," is not a single work but instead a compilation of various treatises and commentaries composed by

Fig. 2.8. Diagram accompanying Viviani's Proposition 5 (detail). *Grandi and Viviani*, Trattato, *14 (detail). RB 708047, The Huntington Library, San Marino, California.*

Galileo, Viviani, and Torricelli, who had passed away more than twenty-five years before the book's publication. The first two sections relate to the so-called Fifth Day of the *Two New Sciences*, a dialogue among the same three interlocutors of the original four days, who now turn to a discussion of the fifth book of Euclid's *Elements*.[86] Galileo's actual dialogue (the "Fifth Day") is the second work in the collection. The first work, titled "Universal science of proportions," is a compilation treating the fifth book of Euclid that draws in part on the demonstrations contained in Galileo's dialogue. Viviani followed these two sections with extracts from Galileo's letters relating to the publication of the *Two New Sciences*, Viviani's own remembrances of Galileo, and various digressions and other opinions of Galileo and Viviani on mathematical topics. The last four sections of the text address Torricelli's unpublished works.

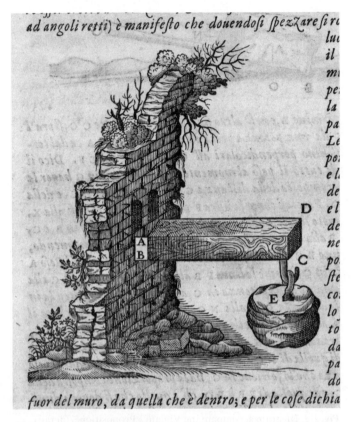

ad angoli retti) è manifesto che douendosi spezzare si ra

Fig. 2.9. Diagram accompanying Galileo's first proposition in Day 2. *Galileo,* Two New Sciences *(1638), 114 (detail). RB 701317, The Huntington Library, San Marino, California.*

Viviani included demonstrations by Torricelli of various propositions contained in the sixth book of Euclid's *Elements*, extracts from Torricelli's correspondence relating to the publication of his *Opera Geometrica,* and Viviani's own additions to the first book of Euclid and reflections on the excellence and utility of geometry.

The contents of this collection suggest that it was produced by following the reading strategies and goals Viviani employed when he annotated the *Two New Sciences*. Most obviously, the book was an editing project that involved seeing into print unpublished treatises composed by Galileo, Torricelli, and Viviani. Parts of the publication were produced by compiling extracts of other writings, especially letters, all related to a specific topic, and gathering and ordering material discussed in the dialogue of Galileo's Fifth Day in the format of an orderly mathematical treatise, the "Universal science of proportions."[87]

A closer examination of the "Universal science of proportions" reveals that its production was predicated on Viviani's application to Galileo's Fifth Day of the reading strategies apparent in his annotated copy of the *Two New Sciences*. The "Universal science of proportions" is structured as a series of propositions contained in the fifth book of Euclid's *Elements*. The definitions, axioms, and demonstrations, however, do not all derive from Euclid but were gathered from a variety of sources, some from Galileo's Fifth Day, others elaborated on by Viviani, and still others drawn from the writings of earlier mathematicians. The treatise is a product of Viviani's reading, excerpting, and modifying demonstrations he found elsewhere, principally in the dialogue that makes up Galileo's Fifth Day. As he explained to the "noble students of geometry," to whom the treatise is addressed, he produced the text by setting himself the task of repairing (*riformare*) and extending (*distendere*) Galileo's own demonstrations.[88] Such activities were precisely those that constituted his response to the *Two New Sciences*, in which he also repaired and extended much of Galileo's text. The "Universal science of proportions" can thus be seen as a more formalized set of Viviani's reading notes than those found in the margins of his copy of the *Two New Sciences*, this time excerpted and elaborated from the Fifth Day.

This relationship is clearly indicated through the presence of printed marginalia alongside the dialogue of the Fifth Day. Intended by Viviani explicitly as a way to help his own readers move between his "Universal science of proportions," Galileo's dialogue, and Euclid's text, these marginalia key the discussion of Galileo's interlocutors to the relevant proposition number in Viviani's "Universal science of proportions" and in Euclid's *Elements*.[89] Thus, Proposition 1 in the "Universal science of proportions" is identified by Viviani as a statement of proposition 4 in book 5 of Euclid's *Elements*. The marginal note in the "Universal science of proportions" reads "Proposition 4 of the Fifth of the *Elements* proved following Galileo."[90] Its demonstration derives from a section of the dialogue in Galileo's Fifth Day, alongside which Viviani has added the corresponding marginal note, "Proposition 1, or the fourth of the Fifth Book of Euclid."[91] Other marginalia in the *Trattato* indicate the sources of Viviani's text. These include, of course, more material from Galileo's Fifth Day; Viviani, for example, took many of his definitions from Galileo's text, noting in the margins that such definitions were "explained in a different way following Galileo."[92] Some material was derived from Viviani's own elaborations, which he indicated in various ways, including in marginal notes such as "Definitions 8 and 9 of the Fifth Book of Euclid explained more amply" and "Second part of Proposition 8 of Book 5 of the *Elements* demonstrated by me."[93] Other definitions, axioms, and demonstrations were culled from the published and unpublished writings of other authors, including Torricelli and Christoph Clavius (1538–1612).[94]

The structure of the entire publication, moreover, reveals Viviani's desire to fold Galileo into an ancient lineage of mathematicians. In his "Universal science of proportions," Viviani augmented his discussion of Euclid's text with a set of propositions and proofs that Euclid's later interpreters had added to his work. He explained to the reader that many of these later commentators had recognized the need for these additional propositions and that he had therefore included them as well.[95] By organizing the treatise in this way, Viviani presented Galileo as one of many commentators on Euclid, alongside such authorities as Pappus of Alexandria (ca. 290–ca. 350), the more recent Campanus of Novara (ca. 1220–1296), and Clavius. Like the writings of these commentators, the demonstrations in the dialogue of Galileo's Fifth Day could be folded back into a sustained discussion of the *Elements*. Viviani thus styled Galileo's Fifth Day as an additional commentary, composed in dialogue form, on Euclid's fifth book.

Viviani portrayed himself and Torricelli as the latest inheritors of this tradition. In seeing the work into print, extracting Galileo's demonstrations into the "Universal science of proportions," and inserting printed marginalia in the Fifth Day, Viviani assumed the role of a commentator on Galileo's commentary on Euclid. Subsequent sections of the treatise contain Torricelli's and Viviani's own additions and alternative demonstrations to Euclid's text. In the *Quinto libro*, then, Viviani placed Galileo firmly in this tradition of mathematical commentators. Galileo and his students were successors in this tradition, not individuals who broke with revered forms of scholarship.

Viviani is often named as a member of the "Galileo school," an individual who understood Galileo's new methods and tried to promote them. The assumption underlying such a designation is that Viviani and this "Galileo school" promoted a type of science that was distinct from that embraced in the sixteenth century and by the wider community in the seventeenth.[96] A closer look at Viviani's reading of the *Two New Sciences* and the intellectual projects that grew from it, however, reveals that his scholarly practices were a continuation of sixteenth-century methods. Like other early modern readers, he relied on a variety of bookish practices, from marginal annotations to the copying of extracts and annotations. His engagement with the content of Galileo's text resembles that of earlier readers of mixed-mathematical texts in mechanics and astronomy. Viviani read Galileo to understand his teacher's mathematical proofs, and he used his annotations for scholarly projects oriented toward explicating Galileo's mathematics. In his *Quinto libro*, moreover, Viviani portrayed Galileo as the latest successor in a long line of distinguished mathematical commentators. These observations suggest a strong continuity in scholarly practices before and after Galileo. Despite Galileo's and even Viviani's insistence on the novelty of Galileo's methods, his last surviving student relied on traditional approaches as he read and worked with Galileo's text.

Viviani's reading of the *Two New Sciences* appears to be largely traditional in another respect too, namely, in its disciplinary specificity. Viviani corresponded with Mersenne and others in the middle of the second Galileo affair, the debate that erupted in the 1640s over Galileo's science of motion precisely because of its ties to Galileo's pro-Copernican outlook. We know, too, that the problem of Galileo's condemnation was a constant and unwelcome presence in Viviani's life. He toiled to preserve papers that presented Galileo in the best possible light and to secure permission for a memorial tomb for his teacher, while also keeping abreast of his contemporaries' biographies of Galileo and depictions of his condemnation. Despite this engagement with these issues of astronomical and cosmological significance, Viviani's marginalia in the *Two New Sciences* offer no indication that he read the book in connection with Galileo's Copernican commitments. Although Viviani referenced outside texts in his marginalia, nowhere did he cite relevant passages from the *Dialogue*. Like early readers of Copernicus who focused on the astronomer's mathematical models and ignored Copernicus's physical hypotheses, Viviani read the *Two New Sciences* on its own, largely as a mathematical text, without noting its cosmological or, for the most part, its physical implications. This disciplinary specificity may have been deliberate. Keen to rehabilitate Galileo's reputation, Viviani may have read the *Two New Sciences* and edited its later additions within the safe confines of Euclidean geometry precisely in order to avoid the suggestion that they touched on more dangerous areas of scholarship.

The evidence of Viviani's reading thus extends the portrait of the ideal reader of Galileo advanced in chapter 1. In Viviani we have a reader who was not focused on the veracity or physical significance of Galileo's claims, but more on Galileo's opinions, conclusions, and mathematical reasoning. Rather than using experiment or mathematics to extend Galileo's conclusions, Viviani applied a variety of older forms of reading strategies aimed at summarizing and extracting the key opinions and demonstrations advanced by Galileo to other usable forms, whether marginalia or separate treatises. Moreover, despite the overwhelming presence of Galileo's Copernican commitments and condemnation in his own scholarly and professional life, Viviani read, or at least gave the appearance of reading, the *Two New Sciences* in discipline-specific ways, treating the text as belonging to the genres of mixed mathematics and geometry.

CHAPTER THREE

Modifying Authoritative Reading
to New Purposes

This chapter moves further geographically to consider Marin Mersenne, a contemporary of Galileo's who expressed a similar affinity to his work as did Baliani and Viviani but whose intellectual and social circles in Paris were far removed from the court and university environments, in which Galileo and his Italian readers operated. It reveals that marginalia could serve not only as a record of engagement with one text but also as a trading zone where different traditions of scholarly practices came together. Mersenne's marginalia show how and why.

Mersenne was a key figure in the Parisian and international intellectual community in the seventeenth century. His rooms at the Minim monastery served as the meeting place for philosophers and mathematicians in Paris, while his prolific correspondence served to connect an international network of philosophers and mathematicians. Through his publications and letters, Mersenne diffused the works of such eminent contemporaries as Thomas Hobbes (1588–1679), Gilles Personne de Roberval (1602–1675), and René Descartes.[1]

These efforts extended to the writings, published and unpublished, of Galileo, and Mersenne has been a central protagonist in accounts of Galileo and the spread of his ideas. Though his earliest remarks about Galileo were less than enthusiastic, by the beginning of the 1630s Mersenne had begun an ardent campaign to promote and to champion Galileo's work on motion and mechanics. In addition to sharing Galileo's ideas with his regular correspondents, Mersenne took it upon himself to communicate them to a wider audience. In 1634 he saw into print a translation of one of Galileo's early treatises on mechanics, then unpublished but circulating in manuscript, accompanied by his expansions and commentaries. His 1634 *Questions théologiques* detailed arguments found in Galileo's *Dialogue*. Five years later, Mersenne published another unauthorized translation of Galileo's work, entitled the

Nouvelles pensées, which offered a summary and commentary in French of Galileo's 1638 *Two New Sciences*.

Galileo's writings shaped Mersenne's views and experimental programs. The *Dialogue* persuaded him that he was wrong in thinking that falling bodies either slow down or move with uniform speeds. In his *Traité des mouvemens*, which appeared in 1633 but is dated 1634, Mersenne endorsed Galileo's odd-number rule. Galileo's work on heavy bodies in motion, contained in both the *Dialogue* and the *Two New Sciences*, also deeply penetrated Mersenne's later writings. His 1636–37 *Harmonie universelle*, on music, begins with a treatise on sound, its nature, and its properties that includes extensive discussions of motion in general. Here, Mersenne drew on Galileo's *Dialogue* explicitly and implicitly to address a variety of problems involving free fall, movement along inclined planes, and the motion of pendulums. In his 1644 *Cogitata Physico-Mathematica* and his 1647 *Novarum Observationum Tomus III Physico-Mathematicarum*, Mersenne discussed Galileo's findings on mechanics and motion more generally. At times, he included specific citations of Galileo's *Two New Sciences*. Mersenne referenced several findings from Days 1 and 2 of the *Two New Sciences* when he discussed the resistance of materials to breaking in the *Cogitata*.[2] In his discussion of ballistics, Mersenne reproduced and corrected the table of projectiles found in Day 4.[3]

In these publications, Mersenne related his own experimental observations. He often supported Galileo's theoretical claims, while taking issue with the specific numbers Galileo reported or the level of detail offered in his textual descriptions. In one passage in his *Nouvelles pensées*, for example, Mersenne complained that Galileo had failed to describe in sufficient detail the equipment Galileo had employed to measure the heaviness of air compared with that of water. According to Mersenne, Galileo had not specified the weight of the flasks he used, the force and correctness of his balances, or even the heaviness of the air he measured. All of these omissions, Mersenne claimed, left the reader with doubts regarding the accuracy of Galileo's reported numbers.[4] Mersenne's interest in verifying Galileo's numerical results also can be seen in the attention he paid to metrology. Beginning in the *Traité*, Mersenne expressed his concern with the specific relationship between the Florentine braccio, which Galileo referenced in his writings, and the Parisian royal inch. In his *Harmonie universelle*, he assumed that the braccio measured 20 royal inches, but he later added a corollary citing Nicolas-Claude Fabri de Peiresc (1580–1637) to correct this value to 21½ royal inches.

Finally, in Rome in 1644 Mersenne found a shop near St. Peter's that sold samples of units adopted in Italy and determined the correct value to be 23 Parisian inches.[5]

Mersenne's enthusiasm was but one instance of a larger show of support for Galileo by scholars residing in France, one fueled both by Galileo's astronomical discoveries and by his condemnation. Mersenne's earliest printed mentions of Galileo referred to his *Starry Messenger* and his sunspot observations, not to his work on mechanics.[6] In the summer of Galileo's trial, Mersenne hurried to assemble and send to Italy his correspondents' responses to the *Dialogue*.[7] Galileo's trial and subsequent house arrest, news of which circulated in the December 1633 edition of Théophraste Renaudot's *Gazette*, prompted sympathy and solidarity from a number of scholars, many of whom were members of Mersenne's circle.[8] The news impelled Elia Diodati, a correspondent of both Mersenne's and Galileo's, to collaborate with Matthias Bernegger to issue a Latin translation of the *Dialogue*, which was published in 1635.[9] Shocked that Galileo, a friend of the pope's, had been condemned, Descartes suppressed publication of his *Le Monde* and treated the heliocentric world system as a hypothesis in later writings. Peiresc attempted to exert his influence to obtain a pardon for Galileo, while Count François de Noailles (1584–1645), Galileo's former student and dedicatee in the *Two New Sciences*, engaged in similar efforts.[10] These supporters were motivated to correspond with Galileo and assist in the dissemination of his work by the prevailing ideals of open communication advocated by the nascent Republic of Letters. They saw themselves as the newest inheritors of ancient Greece and Rome and as protectors of intellectual freedom against papal domination. Their devotion was inspired by Galileo's vocal embrace of classical mathematics and his defense of Copernicanism in the face of opposition by the Catholic Church.[11]

Mersenne's desire to validate Galileo's findings, his attempts to experiment, his focus on the science of motion, and his reading the *Dialogue* and the *Two New Sciences* in tandem makes him the embodiment of Galileo's "ideal" reader presented in chapter 1. He has been central to accounts of Galileo's reception. In his *Thinking with Objects*, Domenico Bertoloni Meli devoted a third of his chapter on the reception of Galileo's mechanics to Mersenne.[12] Paolo Galluzzi identified Mersenne as "the active and able director" of the second Galileo affair and chose Mersenne's death in 1648 as the endpoint of the controversy.[13]

This chapter enriches this picture by taking a closer look at the reading practices Mersenne employed as he carried out these projects. Focusing on two copies of the *Two New Sciences* annotated by Mersenne, it reveals that even as Mersenne claimed to endorse and embrace the experimental and quantitative methods of the New Science, his engagement relied on the textual practices of sixteenth- and seventeenth-century readers. More so than Viviani, Mersenne moved beyond the traditional reading methods he appropriated, clearly eager to employ texts in new ways that reflected his changing conception of what it meant to do natural philosophy. Mersenne's modification of traditional practices, however, both enhances the model of Galileo's "ideal" reader outlined in chapter 1 and nuances the standard account of seventeenth-century scholarly practices. With regard to the former, Mersenne's reading and annotating methods reveal the significant ties he retained to earlier forms of scholarship, which embraced textual comparisons and authorities, as opposed to approaches that relied on experiment or mathematics to validate and extend textual claims. In terms of the latter, the example of Mersenne suggests that textual, scribal practices were ubiquitous in the middle decades of the seventeenth century and not just a feature of descriptive, Baconian sciences, as recent scholarship has often implied.

Diagrammatic and Repositorial Reading in the BIF and BSG Copies

Two surviving copies of the *Two New Sciences* annotated by Mersenne survive. One copy, held at the Bibliothèque de l'Institut de France (BIF), was first identified by Claudio Buccolini and is annotated sparingly.[14] It contains, apart from an ex libris in Mersenne's hand, only three additional annotations. One is an extensive marginal note, the second is a small correction in Galileo's wording, and the third is a diagram inscribed alongside the text in Day 3.[15] Unlike many of Galileo's other readers, Mersenne did not even take the trouble in this copy to correct the printers' errors listed at the back of the book.

A second copy, which is more heavily annotated, is held at the Bibliothèque Sainte-Geneviève (BSG).[16] Of the copy's 314 pages of text, 59 are annotated. Day 1 contains primarily textual notes in the margins and some in-line corrections. The annotations in Days 2, 3, and 4 are predominantly geometrical diagrams; only 7 of the 34 pages annotated in Days 2 through 4 consist of textual notes. The copy is missing three folios at the end, which comprise the printed index and errata list, suggesting that it is one of the surviving first issues of the book. The inside cover contains the trace of a known provenance mark, which links the copy to an important collection of more than a thousand volumes that made up a private library in the seventeenth century. The title page offers an additional indication of provenance, an ex libris of the Abbey of Sainte-Geneviève affixed in the early part of the eighteenth century, indicating that the book entered the holdings of the abbey along with the rest of this private collection before 1750.[17]

Following a series of administrative reforms during the French Revolution and the Bourbon Restoration, the holdings of the library were augmented and put under state control, with the library eventually receiving its modern designation as the Bibliothèque Sainte-Geneviève.[18] Aside from these marks of provenance, the copy contains no other identifying marks. The annotator's identity as Mersenne was confirmed by comparing the marginalia in the BSG copy to known samples of Mersenne's handwriting.[19]

Unlike other early modern scholars, including Erasmus Reinhold, Gabriel Harvey, and Isaac Casaubon, Mersenne left no statement on either the BSG or the BIF copy to explain his motivations or methods of annotation.[20] Examination of the marginalia, however, suggests that both copies derive from the same source and were annotated with comparable methods and for similar purposes.

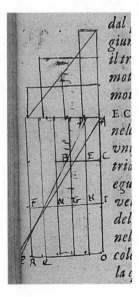

Fig. 3.1. Mersenne's redrawing of a diagram found on the previous page. Two New Sciences (*1638*), *174 (detail).* *Courtesy of the Bibliothèque de l'Institut de France, Paris. Shelfmark 4° M 541.*

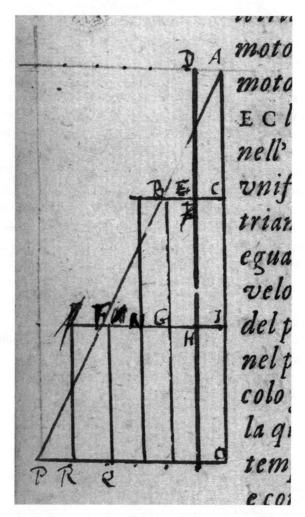

Fig. 3.2. Mersenne's redrawing of the same diagram in the BSG copy as in figure 3.1. Two New Sciences *(1638),* 174 *(detail).* ©Bibliothèque Sainte-Geneviève, Paris. Shelfmark 4 V 585 INV 1338 RES.

Both copies contain diagrammatic annotations made in the same pattern. The one diagrammatic annotation in the BIF copy, in Day 3, is a redrawing by Mersenne of a diagram found on the previous page (fig. 3.1). The manuscript diagram most likely was intended as an aid for following along with Galileo's argument. The BSG copy contains numerous diagrammatic annotations, especially in Days 3 and 4. It even includes a cleaner version of the same diagram found in the BIF copy (fig. 3.2). Other readers worked with printed diagrams in this way. For example, while seeking a printer for the *Two New Sciences* in Bohemia, Giovanni Pieroni had foreseen the difficulties in-line diagrams would pose to readers and had suggested printing all the diagrams on additional sheets at the front or the back of the book.[21] Because the Elseviers chose to use in-line diagrams, other readers, including Viviani, followed Mersenne's strategy and copied Galileo's figures on successive pages.[22]

The textual marginalia in both the BIF and BSG copies also appear to derive from the same source, namely, a letter sent by Descartes to Mersenne in October 1638.[23] Descartes had written in response to Mersenne's request for an assessment of Galileo's recently published *Two New Sciences*. In the letter, Descartes provided a general appraisal of Galileo's approach to philosophy followed by a list of comments and observations made in reference to specific pages in the 1638 edition. Of the thirty-five pages Descartes specifically mentioned in his letter, the comments for all but four of them were copied in some form by Mersenne into the BSG copy. The one textual annotation in the BIF copy is a word-for-word rendition of a statement made by Descartes in his 1638 letter. These textual annotations thus are not Mersenne's responses to the *Two New Sciences* but rather Mersenne's extracts, summaries, and occasionally verbatim transcriptions of Descartes's letter. The BIF copy appears to be an aborted attempt to record Descartes's comments in the style of the BSG copy.

The copying of annotations into multiple copies of the same book was not unheard of in the period. Paul Wittich, for example, owned and annotated at least four copies of Copernicus's *De Revolutionibus*. While few of the annotations in the copies are absolutely identical, many are similar. There are indications that Wittich annotated and corrected the books in tandem, using conclusions he drew in the marginalia of one copy to amend earlier notes he had inscribed elsewhere.[24] Other readers of Copernicus, including Johannes Praetorius (1537–1616), Michael Maestlin (1550–1631), Duncan Liddel (1561–1613), Johannes Broscius (1585–1652), and Erasmus Reinhold, also left to posterity multiple copies of *De Revolutionibus* containing their annotations.[25] The practice of sharing annotations and books was also widespread among readers of Copernicus, leading to the diffusion of specific interpretations of Copernicus that emphasized his technical mathematical models rather than his novel philosophical hypotheses.[26] The antiquary and astronomer Peiresc is also known to have invited other scholars to read and annotate his own books.[27] The humanist teacher of music Heinrich Glarean shared his own annotations with students, who often subsequently copied them into their own books.[28] Mersenne's use of these two copies as repositories for Descartes's opinions can thus be seen as the continuation of established practices.

Orienting Topical Reading Strategies to a Contemporary Conversation

The importance of scribal technologies to Mersenne's intellectual production as revealed in the BIF and BSG copies is supported by other evidence. Mersenne's surviving letters, for example, testify to the fact that he and his correspondents regularly annotated their books and shared both their books and their annotations as part of their intellectual exchange. In response to Mersenne's request for his opinion on Galileo's 1632 *Dialogue*, for example, Descartes explained that he did not have a copy of the book but that he had based his opinion on a quick perusal with the help of Isaac Beeckman (1588–1637). According to Descartes, "M. Beeckman came here on Saturday evening and lent me the book by Galileo. But he took it away with him to Dordrecht this morning; so I have only had it in my hands for thirty hours. I was able to leaf through the whole book."[29] Descartes went on to provide lukewarm praise for Galileo, who philosophized "well enough" on motion and tended to do a better job when he opposed received opinion.

Descartes was again at a disadvantage when Mersenne wrote to him in 1638 requesting comments on the newly published *Two New Sciences*. In his first response, Descartes noted, "Your last letter just contains observations on Galileo's book, to which I cannot reply, because I have not yet seen it; but as soon as it is available for sale, I will look at it, if only to be able to send you my copy with my annotations, if that would be worth doing, or at least to send you my observations."[30] Here Descartes proposed two ways to share his opinions on Galileo's work: he could either send Mersenne his own annotated copy or make an extract of his annotations and send those in place of the book. The final line of the letter indicates that Descartes and Mersenne judged the annotated copy more valuable, though more tedious to examine, than a summary of Descartes's observations on separate sheets. This evaluation might explain why Mersenne then transcribed the substance of Descartes's letter into the margins of the BIF and BSG copies.

While such sharing of annotations was a common period practice, that Descartes wrote and shared marginalia is surprising given his own rhetoric differentiating himself and his methods from those of his contemporaries. In studies of scholarly methods, historians often have singled out Descartes as an individual whose writings and methods explicitly invalidated the theoretical underpinnings of traditional bookish practices, especially the commonplace method.[31]

Surviving papers give a more expansive view of Mersenne's scholarly practices. One especially rich source is Mersenne's personal copy of his own 1636 *Harmonie universelle*.[32] Mersenne and his contemporaries attached great value to these notes. Just as Mersenne had transcribed Descartes's notes for future reference, Mersenne's marginalia in the *Harmonie universelle* were later copied. An unknown scholar, most likely in the late seventeenth century, wrote out many of the annotations Mersenne included in the *Harmonie universelle* on separate sheets.[33] This scholar made careful note of the page and line numbers to which Mersenne's notes referred, likely because he intended to read and study Mersenne's annotations alongside the *Harmonie universelle*.

Mersenne's annotations in the *Harmonie universelle* reveal that he was thinking explicitly about the connections between the *Dialogue* and the *Two New Sciences* as he took notes. In this work, published two years before the *Two New Sciences*, Mersenne took up Galileo's science of motion in reference to the conclusions presented in his 1632 *Dialogue*. Mersenne used his personal copy of the publication as a notebook of sorts for recording observations taken from his reading, and his marginalia contain more than a dozen notes on Galileo's *Two New Sciences* alone. Alongside one section of his text in which Mersenne addressed passages on local motion in Galileo's *Dialogue*, for example, Mersenne noted additional details Galileo had included in his *Two New Sciences* regarding the effects of the medium's resistance on the speed of fall and the fact that heavy bodies in a void should fall at the same speed.[34] In other annotations made in his *Harmonie universelle*, Mersenne referred to his own translation of the *Two New Sciences*, his 1639 *Nouvelles pensées*, and at times to both the original 1638 edition and his own translation.[35]

Mersenne read and took notes on the *Two New Sciences* in tandem with the *Dialogue* because in his reading he tended to focus his attention on a given set of subjects, such as accelerated motion. Mersenne's marginalia in the *Harmonie universelle*, in fact, can be interpreted as a mechanism for facilitating his topical reading strategies, modified from the traditional commonplace method.[36] The "heads" under which Mersenne took notes were the topics in his own book.

Mersenne was not alone in using a published text as a place to record his reading notes. Harvey used his books similarly and intended them to serve as a record of reading much wider than the individual books in which the notes were taken.[37] Similarly, Carl Linnaeus (1707–1778) annotated his own books through successive editions, not with the goal of producing a "perfect" text but in order to store his own reading notes and observations. Contemporaries, including his son Carl and other readers, engaged in the same practice, using Linnaeus's published writings as organized repositories for their notes.[38] Mersenne's topical reading and writing practices are also apparent in his other publications, including his *Nouvelles pensées* and his manuscript "Livre de la nature des sons."[39]

Drawing attention to Mersenne's topical reading allows for a more pointed comparison between his scholarly practices and those associated with traditional Renaissance natural philosophy. Like many sixteenth-century readers, Mersenne read topically, but he put Galileo's *Two New Sciences* in conversation with his and his contemporaries' works, not with Aristotle or standard questions of traditional natural philosophy. None of Mersenne's annotations in his *Harmonie universelle* comment on the relationship between Galileo's text and Aristotle's writings. In his translation of the *Two New Sciences*, Mersenne too minimized the connections between Galileo and Aristotle, only touching in passing, if at all, on Galileo's summaries and discussions of traditional Aristotelian arguments.[40] What stands out from the surviving records of Mersenne's scholarly practice is his desire to put Galileo in dialogue not with the ancient tradition but with the members of his own circle. Mersenne created a world of textual scholarship that privileged the close reading of and commenting on texts but whose points of reference were modern authors, not ancient ones.

Mersenne also went beyond many of his contemporaries in the way he employed marginalia to facilitate new ways of reading. One example from his marginalia reveals Mersenne in the midst of transitioning between a more traditional, descriptive approach to reading and note-taking and a more critical approach. Alongside a section of his *Harmonie universelle* discussing the pitches produced by harpsichord strings of different materials, Mersenne remarked that he had read on page 103 of the 1638 *Two New Sciences* that strings made of gold are lower in pitch because they are limper, not because they are heavier.[41] On the relevant page of the *Two New Sciences*, however, Galileo argued the opposite, that pitch could be raised in three ways: by changing the thickness, the tension, or the length of the string, all of which alter its weight. Galileo went on to claim that the real determinant of pitch is related to the number of vibrations made by the string.

At first glance, Mersenne's annotation appears to be a misreading of Galileo's text. Whereas Galileo claimed that differences in weight altered pitch, Mersenne said specifically that it was not the string's weight that was responsible but its malleability. Closer examination reveals that Mersenne's note derived not from a poor reading of Galileo's text but rather from Mersenne's misunderstanding of his own cryptic annotations in the BSG copy. In his 1638 letter to Mersenne, Descartes noted that Galileo "says that the sound of gold strings is lower than strings of brass, because the gold is heavier; but it is more because it is more pliant. And he errs in saying that the heaviness of a string resists the speed of its movement more than its size."[42] Descartes, offering a correction to Galileo, thus said that the lower pitch of the string was due to its being more malleable, not to its weight. In the margins of the BSG copy, Mersenne had condensed Descartes's comment, only jotting down that "the sound of gold strings, which is lower, is so because they are more pliant, and not because [they are] heavier, and it is false that the heaviness of a body resists the speed of its movement more than its size."[43] Mersenne here copied Descartes's correction of Galileo but omitted his summary of the text. When Mersenne then turned to annotate his *Harmonie universelle*, he drew directly on his marginalia, not on Galileo's text. Since he had not indicated in the margin that the note was a correction to the passage, Mersenne copied the annotation verbatim, beginning, "On page 103 of Galileo's dialogues one finds the reason why the sound of gold strings is lower." Mersenne claimed that Galileo had stated the opposite of what he actually wrote, when in fact the opinion Mersenne assigned to Galileo was Descartes's.

Mersenne may have misread his notes precisely because the marginalia in his *Two New Sciences* departed in particular ways from established note-taking practice. Ann Blair has argued that early modern readers' predominant interest was in flagging passages of interest, either with nonverbal marks or by highlighting keywords, examples, and authorities mentioned in the text.[44] The purpose of these notes was to serve as a guide or rapid-retrieval system for accessing the author's text, not to express the reader's judgment of or corrections to it. It appears that when Mersenne turned to his marginalia in the *Two New Sciences* to copy excerpts into the *Harmonie universelle*, he read his annotations as if they were reading notes in the sense described by Blair. Mersenne interpreted his marginalia as reflections of Galileo's opinions, not corrections to them. Momentarily oblivious to the innovativeness of his and Descartes's note-taking practices, in the margins of his *Harmonie universelle* Mersenne inadvertently assigned to Galileo Descartes's opinion.

Mersenne's textual practices both rely on and depart from those of traditional natural philosophy. While he read Galileo using paper and pen, taking notes on Galileo's arguments, sharing his notes and annotations with correspondents, and discussing Galileo's views in print, Mersenne put Galileo in conversation with his own contemporaries rather than with the textual authorities central to Aristotelian natural philosophy. He used his annotations to record corrections to a text, not merely its argument or subject. Despite these differences, underlying Mersenne's approach is a set of practices—annotating, note-taking according to topical heads, and the sharing and circulation of texts—that points to the way traditional textual technologies were used and transformed for new purposes in the period.

Mersenne's methods in this regard bear similarities to those of other seventeenth-century scholars. His adaptation of the humanist commonplacing method in the margins of his *Harmonie universelle* resembles similar moves by contemporary English virtuosi. Francis Bacon (1561–1626), for example, advocated note-taking under commonplace headings to keep track not of textual excerpts but of literate experience, to assist in memory, thinking, and the eventual consolidation of useful knowledge. John Locke (1632–1704) designed his own "New Method" of commonplacing, which involved a process of simultaneously taking and indexing notes containing both textual extracts and material gathered from observation, experiment, and testimony.[45] All of these readers, Mersenne included, borrowed certain aspects of traditional, humanist note-taking strategies and adapted them to suit new purposes. Whereas Mersenne was interested in relating textual excerpts to his own research agenda, Bacon and Locke were more concerned with adapting the method to accommodate both experiential and textual information.

Mersenne's efforts to exchange ideas via correspondence, to share printed books and annotations, and to continuously add to his *Harmonie universelle* also resemble the scribal practices of John Aubrey, John Evelyn, and John Ray (1627–1705). These English naturalists exchanged, commented on, and extracted notes from Aubrey's two-volume manuscript "The Naturall Historie of Wiltshire." Like Mersenne, these naturalists collaborated through the exchange of papers and specimens with multiple correspondents, rarely limiting themselves to a two-way communication. They approached their work as always unfinished, as did Mersenne, with his continual note-taking and consideration of Galileo's writings. While the practices of these naturalists have been described as inspired by Bacon and specific to descriptive scholarly disciplines aimed at the collection of information, the example of Mersenne reveals their wider applicability to mixed-mathematical subjects.[46] At the same time, the impulse underlying such activities—the notion that a text is fluid and can be improved by putting it in dialogue with other authors and opinions—was a central feature of the Scholastic commentary tradition.

Notes as Material for Later Compositions

Mersenne's note-taking strategies, like Viviani's, shaped the content and aims of his publications. In some instances, Mersenne used his annotations directly in the production of other works. For example, he likely relied on them to compose a letter to Galileo's supporters in Italy in 1643. Written in Latin, this letter was sent on behalf of the "Mathematicians of Paris," an informal group of scholars that Mersenne organized from his rooms at the Minim monastery that has been recognized as a precursor to the French Royal Academy of Sciences.[47]

While scholars have long remarked on the similarities between Descartes's letter of 1638 and Mersenne's letter of 1643, it is clear from the contents of the BSG annotations that they, not Descartes's letter, served as the direct source for Mersenne as he composed his own letter. First, there are more similarities between the 1643 letter and the annotations in the BSG copy than between the 1643 letter and Descartes's letter of 1638. In general, the 1643 letter abbreviates Descartes's objections in the same style as that of the BSG annotations. Consider, for instance, Mersenne's rendering of Descartes's criticism of a passage in Day 1 in which Galileo discusses the scaling of machines (the parallel passages in bold):

Descartes's letter of 11 October 1638   *BSG copy*

He declares what he wants to discuss,
namely, why it is that large machines,
which have exactly the same shape
and are made of the same material as
smaller ones, are weaker than the latter,
and why it is that a child is less seriously
injured by a fall than an adult is, or a
cat than a horse, etc. It seems to me that
there is no difficulty about this, nor any
reason to construct a new science,
because it is obvious that if the force or
resistance of a large machine is to be
exactly proportional to that of a small
one of the same shape, they should not
be made of the same material. Rather,
[it is obvious that] **the larger one must**   **The larger one must be made of a**
be made of a material that is harder   **material that is harder and less**
and less easily broken in proportion to   **easily broken, in proportion to how**
how much greater its shape and weight   **much greater its shape and weight**
are. And **there is as much difference**   **are. There is as much difference**
between a large and a small machine   **between a large and a small machine**
of the same material as between two   **of the same material as between two**
equally large ones of which one is   **equally large ones of which one is**
made of a material much lighter and   **made of a material much lighter and**
harder than the other.[48]   **harder than the other.**[49]

Mersenne ignored the first part of Descartes's commentary, in which he summarized Galileo's approach and dismissed it, but he went on to copy word for word the second half of Descartes's opinion, in which he elaborated on the material in Galileo's text. The 1643 letter follows the BSG copy, not Descartes's letter; it contains nearly the same comment regarding Galileo's discussion of the scaling of machines as does the BSG copy, and it does not repeat the initial summary of Galileo's approach included in Descartes's letter.[50]

Another case in which the 1643 letter parallels the annotations in the BSG copy more so than Descartes's letter concerns Galileo's discussion of bodies falling in a void. In response to this passage, Descartes had written a lengthy comment, noting that "everything he says regarding the swiftness of bodies that descend in the void, etc., is devised without foundation, because he should have determined before what [the body's] heaviness is; and if he had known the truth, he would know that it is nothing in the void."[51] Both the BSG copy and the 1643 letter contain a substantially abbreviated criticism of this same passage. The BSG copy states, "He [says?] that bodies have heaviness in the void, which is not [true]," while the 1643 letter reads, "He supposes that bodies have heaviness in a void on page 73, which is false."[52] In a few instances, all three sources are substantially the same.[53] In only one remark does the 1643 letter seem to follow Descartes's 1638 letter more closely than it does the BSG copy. In reference to page 42 of the 1638 edition, both Descartes's and Mersenne's letters state explicitly that Galileo is lacking in knowledge of optics. The BSG annotation, in contrast, merely criticizes the appropriate passage and makes a reference to Descartes's 1637 *Optics*.[54]

The presence of one annotation in Latin in the BSG copy offers further indication that it served as the source for the 1643 letter. In relation to Galileo's assumptions regarding the horizontal velocity of bodies thrown upwards, Mersenne included the only Latin annotation in the BSG copy. He wrote, "If these things were true, [what was proposed] is true, but he does not prove [them] true."[55] The corresponding passage in Mersenne's 1643 letter, which was composed entirely in Latin, offers the condensed assessment that "if the things he says on page 236 from that section . . . are not true, the entire treatise totters. But he does not prove that they are true."[56] In contrast, Descartes had offered in French a detailed criticism of the passage in his 1638 letter: "[Galileo] adds a further assumption here, which is no more true than those already made, namely, that bodies thrown up in the air travel at the same speed horizontally, but as they fall, their speed increases at a rate which is proportional to twice the distance covered. Given this assumption, we can readily conclude that bodies that are thrown up move along a parabolic path; but since his assumptions are false, his conclusion may well be very far from the truth."[57] The 1643 letter thus resembles the BSG annotation in its brevity and its focus on Galileo's failure to prove his assumptions. Because this is the only annotation written in Latin, it is possible that Mersenne added this annotation later, as he composed the 1643 letter.

The pattern of annotations, furthermore, shows that Mersenne moved from the BSG copy to the 1643 letter, and not the reverse. Whereas all but four of Descartes's comments have a corresponding annotation in the BSG volume, the 1643 letter remarks on substantially fewer pages in Galileo's original publication. Mersenne annotated thirty-three pages of the 1638 edition with textual notes; Descartes commented specifically on thirty-six. In contrast, the 1643 letter contains comments on only fifteen pages.[58] All but two of the fifteen comments in the 1643 letter bear similarities to the comments found in Descartes's letter or Mersenne's annotations, or both; the two comments that are not similar refer to experiments that contradict Galileo's findings, ostensibly carried out by members of Mersenne's circle.[59] It thus seems likely that Mersenne relied on the BSG copy as a repository for reading notes and that these notes, in turn, provided material for later scholarly productions, including the 1643 letter.

Critically Evaluating Galileo Using Multiple Sources

We might also expect Mersenne to have relied on his annotations in the composition of his paraphrase and translation of the *Two New Sciences*, his 1639 *Nouvelles pensées*. Unlike the 1643 letter, however, the *Nouvelles pensées* offers little to suggest that the BSG annotations were instrumental in its composition. Instead, it appears that Mersenne's annotations provided a means for processing Galileo's text, organizing and shaping Mersenne's thinking. It was but one of several textual sources on which Mersenne relied in producing his translation.

This influence is seen most clearly in Mersenne's selection of passages to criticize. Because the *Nouvelles pensées* was intended as an account of the content of Galileo's original corrected of any inaccuracies, Mersenne used his role as editor and translator to communicate his own judgments on Galileo's findings to the reader, either by proclaiming certain propositions as especially worthy "because no one has shown them before" or by using words like *prétendre* (to claim) or *essayer* (to try) to cast doubt on the accuracy of Galileo's demonstrations. Mersenne's annotations—or at the very least the act of reading Galileo alongside Descartes's opinions—shaped this aspect of his translation and paraphrase.

One especially telling example is in Mersenne's rendition of the section of Day 1 in which Galileo offers his solution to the paradox of Aristotle's wheel and uses this solution to account for the phenomena of rarefaction and condensation. The BSG copy contains no fewer than eight criticisms of this section of Galileo's text, all derived from Descartes's 1638 letter.[60] Only four of these critiques made it into the *Nouvelles pensées*. One concerned Galileo's discussion of the myth of the mirrors of Archimedes, which supposedly burned enemy ships by reflecting light. The annotation alongside this passage, derived from Descartes's letter, parallels the comments in the *Nouvelles pensées* but in abbreviated form: "These mirrors of Archimedes are impossible. See page 119 of the *Optics*."[61] In his *Nouvelles pensées*, Mersenne conveyed this criticism through his word choice. In contrast to his typical description of Galileo as showing (*monstre*) or proving (*preuve*), Mersenne described Galileo here as "trying to establish" (*essayer à restablir*) what was said of the mirrors of Archimedes.[62] While the brightness of lights in the night persuaded Galileo that light has a finite velocity, Mersenne declared that its action is accomplished "so suddenly that the eye is not capable of judging" its velocity.[63] He then proposed that readers consult the *Optics* of an "excellent Author" (Descartes) in order to "disabuse themselves of the many imaginings, which harm the sciences more than they aid them."[64]

This reference to Descartes's *Optics* is only one instance of a more general pattern of textual substitutions Mersenne employed in his writing. In his *Nouvelles pensées*, Mersenne often shortened sections of Galileo's proofs, omitted them entirely, or offered numerical examples or alternate proofs in their stead. In place of Galileo's appendix on centers of gravity, for example, Mersenne substituted a discussion of centers of gravity lifted verbatim from a letter written by Descartes, and he placed it in his preface rather than in a supplementary appendix.[65] Mersenne's willingness to mix his own opinions and those of his correspondents with Galileo's extended to his other publications. His translation of Galileo's unpublished treatise on mechanics, *Les méchaniques*, for example, contains his own speculations on Aristotle's wheel.[66] These substitutions are another indication of how Mersenne's topical reading strategies shaped the way he read and wrote. In his scholarship, texts were fluid entities to be parsed and recombined to produce the best final product. Galileo's *Two New Sciences* set the parameters of discussion in the *Nouvelles pensées*, but Mersenne did not feel the need to reproduce it faithfully. Such an attitude resembles that of university professors across Europe, who touted their teaching as courses of Aristotelian natural philosophy, not because they taught a strict philosophy based only on Aristotle's thought but because Aristotle's surviving texts provided the framework and agenda on which classroom teaching was based.

At other times, the criticisms Mersenne offered in his *Nouvelles pensées* did not even correspond to those in the annotations in the BSG copy. In Day 1, Galileo explains rarefaction and condensation through an analogy with Aristotle's wheel, a mechanical paradox contained in the pseudo-Aristotelian *Mechanical Questions* that asked how two concentric circles, constrained to move on two parallel planes, could cover the same distance on the planes in one revolution, given the obvious difference in their circumferences.[67] After arguing that the paradox could be explained by the assumption of infinitely many void spaces contained within the lines traced by the circles, Galileo argues that the process of rarefaction corresponds to the line traced by the smaller wheel as it was carried along by the larger wheel. In this case, an infinite number of atoms were expanded and an unquantifiable number of void spaces interspersed between them. Condensation, on the other hand, could be understood as analogous to the process by which the larger circle traces a line less than the length of its circumference when driven by the rotation of the smaller circle.

In contrast to his treatment of the passage discussing Archimedes's mirror, Mersenne annotated this section of the *Two New Sciences* with a much more detailed note. In the BSG copy, Mersenne, summarizing from Descartes's letter, wrote, "Everything that he [Galileo] says regarding rarefaction and condensation is nothing but a sophism." According to Mersenne, the circle leaves no void parts between its points, but "it is moved only more slowly." Condensation, when it occurs, involves a contraction of the pores and the departure of a part of its subtle matter. Rarefaction involves the entrance of this subtle matter into the pores.[68]

In his *Nouvelles pensées*, Mersenne similarly found fault with Galileo's explanation, but he offered a different explanation. According to Mersenne, Galileo's analogy with the paradox of Aristotle's wheel is not very persuasive, for "if the smaller circle always skips a point of its line without touching it; it follows that it is not continuous and therefore that it is not a line." Mersenne argued that a better explanation of rarefaction and condensation might be one that relied on the motion of bodies, with higher speeds producing greater rarefaction and lower speeds, greater condensation.[69]

These examples suggest that, in contrast to the 1643 letter, the *Nouvelles pensées* exhibits no direct dependence on the BSG annotations. Mersenne did not comment on all the passages he annotated in the BSG copy. Nor, as Mersenne's response to Galileo's treatment of rarefaction and condensation indicates, did Mersenne always base his criticisms in the *Nouvelles pensées* on the annotations inscribed in the BSG copy. We can surmise that instead Mersenne drew on a variety of sources as he composed the translation.

One source apart from the BSG copy that Mersenne no doubt consulted was the published writing of his correspondents, including Descartes's *Optics*. In his *Nouvelles pensées*, Mersenne moved beyond his annotation and even Descartes's letter to describe a specific example from Descartes's text: "And the excellent author who makes us imagine the extension of light by the example of a stick, which shakes what it touches at the same time that it is pressed, shows us the difficulties of the extension or of the instantaneous movement of light . . . and if one has the slightest difficulty in the world understanding what he teaches about light, which is made by a straight movement . . . he will address these concerns for those who request it."[70] The example of the "stick" is a reference to the first discourse of the *Optics*, in which Descartes attempted to explain the nature of light's movement. In this section, Descartes described the motion of the stick of a blind man, which he argued "will prevent you from finding it strange at first that this light can extend its rays in an instant from the sun to us; for you know that the action with which we move one of the ends of a stick must thus be transmitted in an instant to the other end, and that it would have to go from the earth to the heavens in the same manner, although it would have more distance to travel there than it has here."[71] Descartes did not directly address whether light moves with an infinite or finite velocity. Instead, his example emphasized the rapidity of light's motion rather than its finiteness. Mersenne referred readers to Descartes's recent publication not to refute directly Galileo's claim but to offer a wider discussion for his readers. In doing so, Mersenne demonstrated once again his topical approach to reading, in which his own readers were told to consult a recent work addressing a similar subject.

Mersenne's rendition of Galileo's arguments was also likely influenced by more personal exchanges. It is of note that three of the four passages Mersenne criticized in this section of Galileo's text also were commented on in his 1643 letter.[72] Here, of course, because the *Nouvelles pensées* was written first, it is unclear whether Mersenne's doubts, which he expressed in his translation, colored the letter sent on behalf of the Parisian mathematicians or whether Mersenne's conversations with the group shaped both the translation and the letter.

Epistolary exchanges also likely directed Mersenne's thoughts as he composed the *Nouvelles pensées*. His conviction that Galileo's statements about the speed of light were incorrect, for example, may have stemmed from a series of letters exchanged among his correspondents. Descartes had written to Isaac Beeckman on 22 August 1634 critiquing the premise of an experiment that very much resembled that proposed by Galileo.[73] While Mersenne might not have seen this specific letter, he did exchange missives with his correspondents on the nature of light.[74] In one such letter, written in May

1638, Descartes specifically clarified for Mersenne that his use of the word *instant* in his *Optics* to describe the movement of light did not imply that light moved with infinite speed.[75]

Similarly, the phenomena of rarefaction and condensation—and their relationship to the paradox of Aristotle's wheel—had long engaged Mersenne in print and in manuscript correspondence.[76] In the preface to his *Les mécha-niques,* Mersenne discounted the use of Aristotle's wheel to explain rarefaction and condensation.[77] Epistolary exchanges also likely prompted Mersenne's criticism. In a letter of 28 April 1638, Mersenne made comments he would later echo in his *Nouvelles pensées,* asking Descartes whether any rolling ball can actually be said to form a continuous line on a plane, since in rolling, the plane should only be touched by individual points.[78]

These examples indicate the range of material with which Mersenne worked to compose his *Nouvelles pensées*: Galileo's text, his own annotations of it, the letters of contemporaries, and other printed works. The annotations in the BSG copy served Mersenne as a repository of Descartes's opinions on Galileo's text and, at times, a store of material for subsequent annotations and compositions. They are not a record of Mersenne's unmediated response to the *Two New Sciences* but a product shaped by and intended to facilitate early modern textual practices, in much the same fashion as were Viviani's marginalia.

From Descriptive to Critical Reading

A central characteristic of Mersenne's annotating and editing practice was his desire to correct and improve Galileo's text. In this aspect of his reading, he shared something in common with his contemporary Descartes, whose letter to Mersenne, while containing brief summaries of Galileo's points, focused much more on areas in which Descartes believed Galileo had gone astray. This type of critical reading is one that on the whole distinguishes Mersenne and Descartes from earlier, sixteenth-century readers of natural philosophy. Ann Blair has described how Isaac Casaubon used his marginalia to complain about Jean Bodin's treatment of useless, boring, or trivial points in his *Theatrum.* Yet readers of Jean Bodin's compendium largely refrained from criticizing Bodin for being wrong on specific philosophical, philological, or chronological claims. For the most part, Bodin's readers noted his main arguments without comment, reading the text closely to extract from it philosophical judgments, bits of information, and pious arguments. Blair has argued that this focus on textual arguments for the purposes of compilation and reuse was a feature of bookish natural philosophy before the transformations associated with the seventeenth-century New Science.[79] It is useful to draw a distinction here between the descriptive approach of Bodin's readers and the more critical aims of Descartes and Mersenne.

Yet, we should not be too quick to label Mersenne and Descartes as reading revolutionaries. Other instances of sixteenth-century readers who read critically suggest that the transition from descriptive to critical reading was neither one of mere chronology nor associated exclusively with those who embraced the New Science. William Sherman has described the response of an unknown reader to Francis Bacon's 1605 *Advancement of Learning* in a copy held at the Folger Library. Among other annotated copies inspected, this copy stood out for the "resistance" exhibited by its reader, who annotated Bacon's text with critical comments somewhat akin to those Descartes applied to the *Two New Sciences*. Like Descartes, who censured Galileo's approach to philosophy, this reader found fault with some of Bacon's aims, most notably his decision not to cover matters of divinity. This unknown reader was especially enraged by Bacon's treatment of his sources. He accused Bacon of ignoring entire treatises by Aristotle, as well as by later writers, both ancient and modern. He also took Bacon to task for his opinion that fiction was a better teacher than history.[80]

An even more relevant example is the response of Guidobaldo del Monte to two treatises on the balance, one written by his contemporary Giovanni Battista Benedetti and the other by Jordanus of Nemore. In his marginalia in both books, del Monte exhibited reading strategies drawn from the mixed-mathematical tradition similar to those employed by Viviani. However, unlike Viviani, del Monte reacted pointedly and critically to various conclusions reached by the authors. Del Monte, for example, noted that the results of many of Jordanus's theorems were false.[81] His marginalia in Benedetti's book were even more extensive, indicating his overall impression that the assumptions underlying Benedetti's proofs were incorrect, leading to false or problematic claims.[82] One aspect of Benedetti's account that irked del Monte tremendously was his continuing insistence on the novelty of his approach. Again and again, Benedetti proclaimed that he was the first to treat a particular subject or the first to treat it in depth. In response, del Monte noted repeatedly that he himself had written on the same subject or reminded Benedetti that earlier authors, including Aristotle, deserved credit.[83]

What is similar about the three texts that engendered these pointedly critical readings is their authors' insistence on novelty. Bacon, Benedetti, and Galileo all vociferously claimed to be doing something new. At least two of the three critical readers who reacted to their claims, moreover, had a stake in them, for del Monte and Descartes also purported to be carrying out similar work on their own. These observations suggest that what might explain the turn from descriptive to critical reading is authors' insistence on their novelty coupled with readers who had a vested interest in their own aspirations to innovation. This explanation accounts for the response of del Monte and of Descartes. It also lends further support to the identification of pseudo-Baliani, in chapter 1, as Baliani himself. As the author of his own treatise on local motion, Baliani would have read Galileo's *Two New Sciences* with interest and a critical gaze, comparing what Galileo wrote not with what authors of other books had written but with Baliani's own ideas.

What separates the descriptive practices noted by Blair from those of Descartes, del Monte, and Baliani is not so much disciplinary—natural philosophy as opposed to mixed mathematics—as the conception of the intellectual project. Bodin and his readers made few claims to novel ideas. Their enterprise was the collation of opinions and facts to consider established questions. The distinctiveness of individual scholarly projects lay in the way the questions and opinions were composed and the range of works and examples culled. When one approached a contemporary text in this style of scholarship, one read to fortify one's own arsenal of textual examples; reading and finding diverse opinions only strengthened one's own scholarship, even if the author expressed an argument with which one disagreed. In the case of scholars staking out claims of novelty, however, this collaborative model disintegrated. A reader could take issue with an author who claimed to be an innovator; and readers, judging from the examples of Descartes, del Monte, and pseudo-Baliani, did, especially if they thought the author had overstated his or her originality or taken the wrong approach.

Mersenne's methods do not fit this model so exactly. He responded critically to Galileo in his translations, publications, and letters, and he wrote on topics similar to those Galileo addressed. Yet in his marginalia he noted not his own critical comments but those of a contemporary. Mersenne's turn to the responses of contemporaries may signal a lacuna in the historical record; perhaps he did annotate the *Two New Sciences* extensively with critical comments but the copy has not survived. Another likely explanation is that Mersenne's copy represents a hybrid approach, one that drew on established practices of reading in conversation with textual authorities but recognized contemporary criticism of them. Mersenne retained the practices of a bookish style of philosophizing, putting texts into conversation with the opinions of others and relying on strategies such as commonplacing and topical reading, but he looked to contemporaries, not ancient sources, for this conversation.

Mersenne sought to understand the *Two New Sciences* by taking notes, annotating the book, and putting it in dialogue with other writings. He viewed the *Two New Sciences* as a fluid text that could be improved upon by the comments and additions of others. That Mersenne would have employed such methods makes sense, as he himself had been educated in and was well versed in the "learning of the schools" and thus was familiar with the content and practices of Scholastic and humanist pedagogy.[84] Noting this aspect of Mersenne's approach to the *Two New Sciences* broadens and enriches the portrayal of the "ideal" reader described in chapter 1. Mersenne's case reveals the central role played by textual practices and comparisons in the reception of Galileo's work on mechanics. Galileo's text and readers' writing about it were not mere conduits of experimental trials, mathematical reasoning, and philosophical logic; these textual creations were key intellectual products in their own right, shaping the way readers responded to Galileo's ideas.

CHAPTER FOUR

An Annotated Book of Many Uses

What happens when we move to a distant cultural setting, one far from the world of Galileo and his students and outside the realm typically studied by historians of Galileo's mechanics? Two additional surviving exemplars of the *Two New Sciences* allow us to do exactly that. They contain superb records of how their annotators, readers not often mentioned in connection with Galileo and his work, responded to his text. One copy's reader has been previously identified and mentioned briefly by J. A. Bennett as Christopher Wren, the architect responsible for the reconstruction of St. Paul's Cathedral in London following the Great London Fire of 1666, member and president of the early Royal Society, and Savilian professor of astronomy at Oxford from 1661 to 1673.[1] The second copy contains the annotations of Seth Ward, a founding member of the Royal Society, Savilian professor of astronomy from 1649 to 1660, and later bishop of Salisbury. Both Ward and Wren annotated Galileo's book in Latin, and their annotated copies were donated to the Savilian Library at Oxford.[2] In 1619 Henry Savile created two Oxford professorships—the Savilian chairs—in astronomy and geometry and donated his own books to form a library for use by the appointed professors. Later professors added their own volumes to the collection, which now numbers around 1,180 volumes.

The names of Ward and Wren are synonymous today with novel currents in seventeenth-century scholarship. As founding members of the Royal Society and early presidents, they were closely involved with experimental approaches to the study of nature. Both, too, served as Savilian professors of astronomy at Oxford, where they promoted the newly developing techniques of algebra, in addition to contributing to the experimental efforts of the Oxford Philosophical Club, an important precursor to the Royal Society. Their own scholarly endeavors often focused on mechanics and astronomy, and they applied mathematics and experiment to the study of nature in ways that resemble Galileo's. Ward, for example, published astronomical treatises exploring modifications to Kepler's laws, while Wren undertook studies of the lunar surface, Saturn's rings, and the mechanisms of collision, often invoking Galileo as his inspiration.[3]

All of these features of Ward's and Wren's intellectual interests suggest that the two might have read the *Two New Sciences* with the types of approaches characteristic of the "ideal" reader in chapter 1. This chapter shows instead that their reading strategies depart from this expected trajectory. Though both were active contributors to seventeenth-century work in astronomy, they read the *Two New Sciences* with seemingly little attention to its relationship to the *Dialogue* and to questions of cosmological import. Like Viviani, Ward and Wren largely did not attempt to validate or extend Galileo's findings. Instead they read for a variety of purposes and drew on a plethora of reading strategies, some in keeping with sixteenth-century practices and others new adaptations of older approaches. For Ward and Wren, Galileo's book was less a text of the "New Science" and more a malleable tool for multiple scholarly projects.

Two Nearly Identical Annotated Copies

The Savile collection contains two copies of Galileo's final published work. One (shelfmark Savile Bb.13) is a copy of the first edition of the *Two New Sciences*, while the other (shelfmark Savile A.19) is a copy of the book contained in the first edition of Galileo's collected works, his 1655–56 *Opere*, published in Bologna. Both volumes are heavily annotated throughout, with extensive marginalia and notes also written on the inside covers and flyleaves, as well as on additional sheets inserted into the books.

The annotations in the two volumes are nearly identical. Furthermore, it is clear that those found in the A.19, "Wren" volume (1655–56 *Opere*) were copied from those found in the Bb.13, "Ward" volume. The inserted pages often contain references to page numbers. In the latter edition (fig. 4.1), these references are all clean, but in the former edition (fig. 4.2), they are almost without exception crossed out. At one point, Wren accidentally wrote the page number that corresponds to the 1638 reader's notes (137) and subsequently corrected it (101) (fig. 4.2).

It is the marginal annotations corresponding to these additional sheets that indicate that the annotator of the 1638 copy is Ward. The 1638 reader referenced these additional sheets with marginal annotations such as "See the empty page at the beginning of the book for another demonstration."[4] The

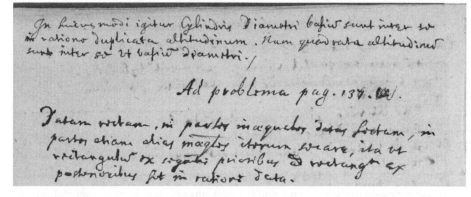

Fig. 4.1. Ward's page-number references. *Savile Bb.13, fly 3v (detail). Courtesy of The Bodleian Libraries, The University of Oxford.*

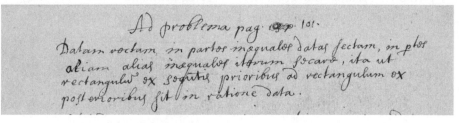

Fig. 4.2. Wren's page-number references. *Savile A.19, slip 2v between pp. 88 and 89 (detail). Courtesy of The Bodleian Libraries, The University of Oxford.*

corresponding annotations of the 1655–56 edition consistently cite not the additional sheets inserted in the book, which are identical to those found in the 1638 edition, but instead the work of "D. Ward," a reference to Dr. Seth Ward.[5] These details suggest that Wren copied his annotations from the 1638 edition and that the marginalia in that earlier edition were written by Seth Ward.

Nothing in the two copies signals when the annotations were composed. Ward's copy was gifted to the Savilian Library, so it is possible that Ward annotated it while he served as Savilian professor from 1649 to 1660. It is also plausible that Ward had the book with him when he arrived in Oxford. We will see, moreover, that references to experiments in the copy indicate that Ward read the text with methods similar to those he ascribed to the Oxford Philosophical Club in the early 1650s. Wren's annotations, contained in the 1655–56 edition of Galileo's *Opere*, were no doubt transcribed after 1655, but it might be that he copied Ward's annotations into another book or on loose sheets earlier and then recopied them into the 1655–56 edition after purchasing it. If this was the case, Wren may have copied Ward's annotations as early as 1650, when he first joined the scholarly community at Oxford as a gentleman commoner at Wadham College, where Ward was then living.[6]

A Record of Experiments to Verify

That the *Two New Sciences* inspired seventeenth-century experimental programs has long been assumed in the historiography. It is often asserted, for example, that Galileo's *Two New Sciences* and other writings served as an impetus for the experiments performed by the Accademia del Cimento in Florence and that these experiments, in turn, served as a model for the Royal Society's experimental program.[7] The *Two New Sciences* is also assumed to have prompted individual readers to undertake programs of experimentation. Galileo's speculations on the void and his observation that there was a maximum height to which water could be pumped led Evangelista Torricelli and others to carry out further experiments on the behavior of mercury and other liquids in inverted tubes. Galileo's student Benedetto Castelli and the Jesuit Paolo Casati (1617–1707) attempted to apply Galileo's quantitative approach to the subject of hydrostatics.[8] In the second half of the seventeenth century, Christiaan Huygens took up many of the problems posed by Galileo and his students in the realms of astronomy and mechanics. He approached them by employing methods, including experimentation and mathematization, that were central to Galileo's own approach.[9] There is a subtle distinction between these assertions and the assumption that the "ideal" reader of Galileo turned to experimentation to verify Galileo's findings. The former assume that readers of the *Two New Sciences* interpreted Galileo's findings in the context of this new turn to experiment. The latter argues that experiment was a natural approach to verify and extend Galileo's findings. This section and the next tease out the variety of ways that two of Galileo's readers employed the textual practices of annotation, commonplacing, and note-taking in relation to these two modes of relating Galileo's book to experiments.

One extensive annotation made by Ward and copied by Wren reveals that at times they responded to Galileo's text in ways similar to those of the "ideal" reader presented in chapter 1, for whom experiment offered the primary means of verifying Galileo's textual claims. This response was prompted by a passage in Day 1 in which Galileo details a method for testing the void. In this passage, Salviati describes a device that he claims allows one to determine by how much force nature prohibits a void (see fig. 2.7). Galileo's proposal is to fill the bucket with weight to force the stopper EFGH to separate from the cylinder; the amount of weight required to do so would be a measure of the strength of the void. In chapter 2 we saw that Viviani approached this passage in his *Trattato* by offering a commentary on it. He responded to the passage by modifying the device and its image and thereby repurposing it to illustrate the principle of a solid's resistance taken absolutely. Ward's annotation, in contrast, focuses on the details and implications of Galileo's proposed contraption, an approach that speaks to his interest in how experiments described in texts could subsequently be carried out and interpreted. Ward began by noting a fundamental assumption underlying the experiment that Galileo never mentioned, namely, that the parts of the axis IK must adhere more closely to themselves than to the surface of the cylinder EF. Otherwise, he noted in the margins, as the stopper was pulled down by the weight, the axis would break, leaving the inner cylinder contiguous with the water, without any empty space.[10] Ward went on to make a somewhat puzzling statement. He claimed that if the inner stopper were pulled from the surface of the water without the iron rod breaking, "were it not for the last experiment," it would be clear that there was another cause of the coherence of the parts in the iron axis.[11] That is, Ward explained, Galileo's attribution of the coherence of parts to nature's abhorrence of the void may not explain the coherence of all matter. Ward then insisted that there could be "another cause of the continual coherence of parts beyond the explanation of the void." To discover the true explanation, Ward wrote, "many experiments" would be useful.[12]

What Ward meant by "the last experiment" is far from clear. It might refer to Galileo's textual description. Alternatively, it may be that Ward carried out his own set of similar experiments inspired by this passage and that he annotated Galileo's text in light of what he found. In both cases, Ward's final note is a direct exhortation to carry out additional experiments in order to understand the cause of the coherence of bodies. Ward's annotations indicate that in his view, the proper thing to do after reading and annotating this passage was to carry out more experiments. This impulse aligns Ward (and Wren) with the responses of the "ideal" reader described in chapter 1.

A Source for New Experiments

Other annotations made by Ward and copied by Wren suggest an alternative way these readers conceptualized the relationship between text and experiment. In particular, they indicate that Ward and Wren employed the *Two New Sciences* as part of a collaborative, Baconian note-taking scheme for the culling of ideas for new experiments, one associated with Interregnum proposals for the reformation of knowledge. In a 1652 letter to Sir Justinian Isham (1610–1675), Ward described the working methods of the Oxford Philosophical Club, a group of scholars at Oxford that included John Wilkins (1614–1672), Robert Boyle (1627–1691), Thomas Willis (1621–1675), Jonathan Goddard (1617–1675), and John Wallis (1616–1703), all of whom later became fellows of the Royal Society.

Ward noted that the club had about thirty members, each of whom takes a portion of "all or most of the heads of naturall philosophy & mixt mathematics" and goes through them, "collecting onely an history of the phenomena out of such authors as we have in our library and sometimes trieing experiments as we had occasion and opportunity."[13] Ward continued:

> Our first businesse is to gather together such things as are already discovered and to make a booke with a generall index of them, then to have a collection of those wch are still inquirenda and according to our opportunityes to make inquisitive experiments, the end is that out of a sufficient number of such experiments, the way of nature in workeing may be discovered, but because (not knoweing what others have done before us) we may probably spend our labour upon that wch is already done, we have conceived it requisite to examine all the bookes of our public library . . . and to make a catalogue or index of the matters and that very particularly in philosophy physic mathematics.[14]

The "public library" to which Ward refers was the Bodleian Library of the University of Oxford, which was founded in 1598 by Thomas Bodley (1545–1613), a one-time fellow of Merton College and a former envoy to the Netherlands, as a public, broadly based academic library. The classificatory project proposed by Ward and the other fellows of the Oxford Philosophical Club was in keeping with the aims of other Oxford scholars involved with the Bodleian in the 1650s. Less than three weeks following Ward's letter to Isham, Gerard Langbaine (1609–1658), the provost of Queen's College, who was responsible for cataloguing the library's Greek manuscripts from 1643 to

1649, proposed a similar plan to survey the contents of the Bodleian and "make a perfect Catalogue of all the Books according to their severall subjects in severall kinds." Langbaine's working papers show that he had developed a scheme for dividing the library's collections into twenty parts, to be allocated to twenty helpers, many of whom were members of the Oxford Philosophical Club.[15]

What Ward described in his letter, while imbued with the same cataloguing and organizing impulses as Langbaine's scheme, was a method of textual commonplacing and indexing carried out for the purposes of experimentation. Ward told Isham that the first task of the Oxford Philosophical Club was to take commonplace headings pertinent to the fields of natural philosophy and mathematics and divide them up among the members. Each member was then to go through the books in his possession and then those in the Bodleian, seeking descriptions of phenomena in the books that fit under the headings. This collection of commonplaces would include "those [things] wch are still inquirenda and according to our opportunityes to make inquisitive experiments." That is, of the lists of items yet to be found, Ward wrote, it would be prudent, when the opportunity arose, to make inquisitive experiments about them.[16]

In the 1650s, the Oxford Philosophical Club was one of many informal groups across the country whose members were influenced by calls to reform English society through spiritual and educational means. Centered at Oxford's Wadham College, the group met weekly on Thursdays, when they presented experiments and discussed their findings.[17] These Interregnum virtuosi took up a number of projects and questions, from the quest for an artificial, philosophical language to scientific and biblical study of the history of the Earth.[18] They were guided by Oxford's tradition of independent mathematical research, new calls to embrace experiment, and Francis Bacon's exhortations to collect information without regard to university learning. Like many English scholars of the 1640s and 1650s, they were inspired by their vision of Adam as the first seeker of natural knowledge, whose innocence and lack of prior commitments allowed him direct access to nature.[19]

These scholars modified traditional reading practices as a means to achieve their spiritual and scholarly goals, and they took Bacon's proposals as a model. One of the key features of Bacon's reforming program was a reliance on note-taking via commonplace headings for the purpose of collating already available information and subsequently collecting new material.[20]

Boyle, a key figure in the Oxford Philosophical Club from the mid-1650s, for example, was an avid note-taker whose surviving papers contain a mix of excerpts copied from books, records of experimental trials, observations, and reports from others.[21] Though chronologically later than the annotations in the Ward and Wren volumes, the early Royal Society also emphasized reading as part of its experimental program. Reading in the Royal Society was a collaborative, multistep endeavor that involved, among other activities, the "perusal" of a work by two fellows, who would read, translate, and abstract it before reporting back to the larger body.[22]

Ward's copy of the *Two New Sciences* is littered with annotations, copied by Wren, which suggest that at times he read the book following the procedure he described for Isham. Many of his annotations specify when Galileo provides a method (*modus*) for carrying out different experimental procedures. Ward noted, for example, Galileo's method for "making gold filaments," his "very accurate method for weighing water," his "method for investigating the ratio of the velocities of heavy bodies in the same and different media," his "method for investigating the heaviness of the air," his description of three methods "for raising the pitch of a chord," his method "for tracing a parabolic line," and the method he describes of measuring time.[23] Given Ward's description of his reading habits to Isham, it is probable that many of these annotations were made by him as part of his collation of things yet to be found and requiring "inquisitive experiments."

Members of the Oxford Philosophical Club may have annotated other books in this way. One potential specimen is a copy of William Davison's alchemical text titled *Philosophia Pyrotechnica seu Curriculus Chymiatricus*. Like Ward's copy of the *Two New Sciences*, this copy of Davison's *Philosophia Pyrotechnica* was also donated to the Savilian Library; its shelfmark indicates that it was donated by either Savile (and thus accessible to Ward) or possibly Ward himself.[24] Whereas Ward annotated the *Two New Sciences* using a variety of approaches, Davison's text is annotated primarily using two styles. At times the annotator noted the topic discussed in the text, indicating in the margin the author's definitions of such terms as *filtration* and *distillation*.[25] The majority of the annotations, however, are notes in the *modus* style, which Ward applied to Galileo's descriptions of procedures and experiments. The annotator of the Davison volume, for example, noted in the margin a "method for coaxing out waters and extracts from succulent things."[26] Moreover, there are indications that the Davison copy was annotated by more than one person.[27] If the copy was annotated by more than one individual, that might suggest the type of communal note-taking project that Ward's letter to Isham implies.

It is possible that annotations such as these translated directly into programs of experimentation, as Ward's description indicates. During the course of their researches, the Oxford group assembled a variety of instruments and apparatus.[28] Best known is their construction of an astronomical observatory and a chemical laboratory, both centered at Wadham College, but the group also alluded to its experimental efforts in the areas of optics, microscopy, and mechanics. In his 1654 *Vindiciae Academiarum*, for example, Ward described a "reall designe," promoted by Wilkins, to erect a "Magneticall, Mechanicall and Optick Schoole, furnished with the best instruments, and Adapted for the most usefull experiments in all those faculties."[29]

The records of the early Royal Society provide some indication of the relationship between Ward's annotations and subsequent experiments prompted by them. Any conclusions in this regard must be tentative, because despite obvious continuities between the two societies, historians have emphasized the tenuous nature of these links.[30] The more detailed records of the activities of the early Society, however, can provide some indication of the interests of the members of the Oxford club. They suggest that whether or not his reading directly inspired actual experimental trials, Ward read the *Two New Sciences* and annotated experimental methods that he and his contemporaries were interested in employing.

In some cases, the correspondence between the methods proposed by Galileo and annotated by Ward and those carried out in the meetings of the Royal Society is very general. For example, Ward noted in the margins of Day 1 that Galileo provides "a very accurate method for weighing waters." In this section, Galileo describes how one could use wax balls of different specific weights to determine the relative heaviness of water and other liquids by testing whether the balls floated or sunk in the liquids.[31] While it seems that the Royal Society never carried out this specific experiment, members did weigh and compare the weights of water and other liquids in the context of other trials.[32] Similarly, Ward and Wren included the annotation "means of measuring time" alongside Galileo's description of the water clock he employed to measure the time in which balls rolled down an inclined plane.[33] We know that members of the Royal Society carried out multiple experiments that involved measuring the time it took bodies to fall, including Robert

Hooke's (1635–1703) experiments dropping heavy bodies off the top of St. Paul's.[34] Moreover, there were occasions when members of the Royal Society attempted to devise new means of measuring the time of especially quick motions, such as in August 1664, when they undertook to measure the velocity of a bullet.[35]

In some instances, there is a stronger parallel between Galileo's procedures as annotated by Ward and subsequent experiments carried out in meetings of the Royal Society. One example is found in Ward's annotations to a passage in Day 1 in which Galileo argues, contrary to Aristotle, that the relative velocities of movables depend on the extent to which the heaviness of the medium detracts from the heaviness of the movable. Alongside Galileo's text, Ward included the note, "A method for investigating the ratio of the velocities of heavy bodies in the same and different media."[36] He signaled his interest in the passage by noting many of Galileo's conclusions, from the statement that "the velocity of descent does not depend on the heaviness of the movable" to the fact that "the same movable resists more the motion of the faster-moving movable."[37] The attention Ward paid to these sections can be seen in his extensive underlining of Galileo's text throughout these pages (fig. 4.3).

Ward also was attentive to the method and sample calculations Galileo proposed for determining the relative speeds with which objects fall in different media. His marginalia show that he followed along with Galileo in making these calculations (see fig. 4.3). In the middle of the lefthand page (p. 76), Ward noted Galileo's assumption that lead ("Pl") was 10,000 times as heavy as air, while ebony ("Eb") was only 1,000 times as heavy; this second proportion for ebony was represented by Galileo and then Ward as a ratio of 10,000 times to 10 times the weight of air. Galileo went on to conclude that the retarding effect of the air meant that lead would lose speed in 1 part in 10,000 in air, while ebony would lose 10 parts in 10,000. That is, if the elevation from which the bodies started was divided into 10,000 parts, the lead would reach the ground by as much as 10 (or at least 9) of the parts before the ebony. Translating these parts into actual units, Galileo declared that a lead ball would outstrip an ebony one by less than 4 inches in a fall from a tower of 200 cubits.

Ward followed Galileo in working through additional examples. In the first, Galileo assessed the fall of ebony compared with the fall of a bladder inflated with air. He began, as before, by comparing the weights of the two

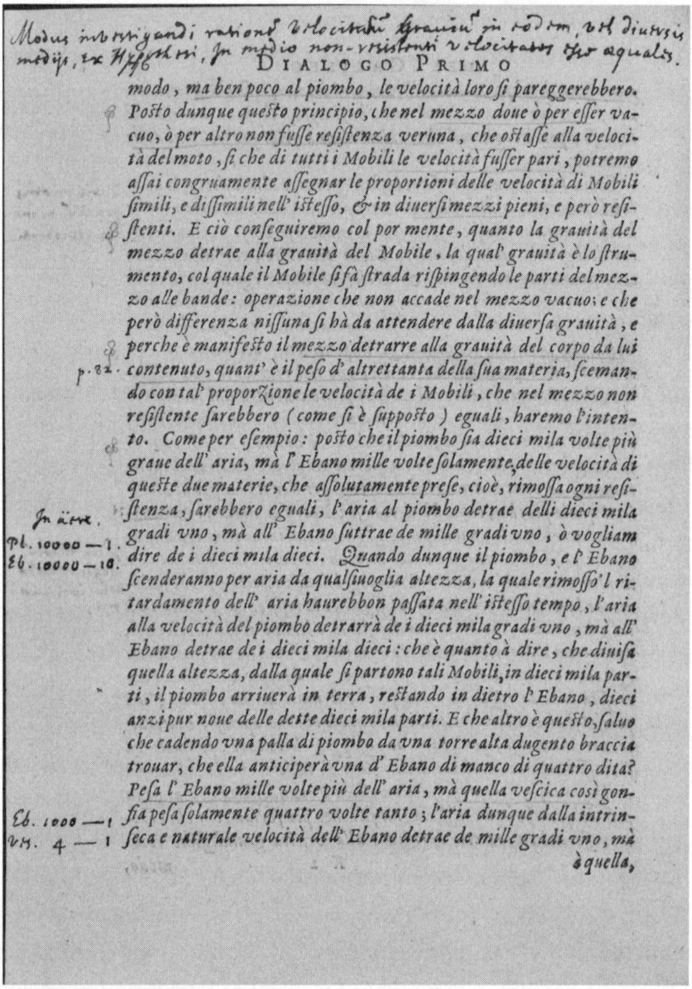

Modus investigandi rationes velocitatum grauium in iisdem, vel diuersis medijs, ex Hypothesi, In medio non-resistenti velocitates sint aequales.

DIALOGO PRIMO

modo, ma ben poco al piombo, le velocità loro si pareggerebbero.
Posto dunque questo principio, che nel mezzo doue ò per esser vacuo, ò per altro non fusse resistenza veruna, che ostasse alla velocità del moto, si che di tutti i Mobili le velocità fusser pari, potremo
assai congruamente assegnar le proportioni delle velocità di Mobili
simili, e dissimili nell'istesso, & in diuersi mezzi pieni, e però resi-
stenti. E ciò conseguiremo col por mente, quanto la grauità del
mezzo detrae alla grauità del Mobile, la qual grauità è lo stru-
mento, col quale il Mobile si fa strada rispingendo le parti del mezzo alle bande: operazione che non accade nel mezzo vacuo: e che
però differenza nissuna si hà da attendere dalla diuersa grauità, e
perche e manifesto il mezzo detrarre alla grauità del corpo da lui
contenuto, quant' è il peso d' altrettanta della sua materia, sceman-
do con tal proportione le velocità de i Mobili, che nel mezzo non
resistente sarebbero (come si è supposto) eguali, haremo l'inten-
to. Come per esempio: posto che il piombo sia dieci mila volte più
graue dell' aria, mà l' Ebano mille volte solamente, delle velocità di
queste due materie, che assolutamente prese, cioè, rimossa ogni resistenza, sarebbero eguali, l' aria al piombo detrae delli dieci mila
gradi vno, mà all' Ebano suttrae de mille gradi vno, ò vogliam
dire de i dieci mila dieci. Quando dunque il piombo, e l' Ebano
scenderanno per aria da qualsiuoglia altezza, la quale rimosso 'l ri-
tardamento dell' aria haurebbon passata nell' istesso tempo, l' aria
alla velocità del piombo detrarrà de i dieci mila gradi vno, mà all'
Ebano detrae de i dieci mila dieci: che è quanto à dire, che diuisa
quella altezza, dalla quale si partono tali Mobili, in dieci mila parti, il piombo arriuerà in terra, restando in dietro l' Ebano, dieci
anzi pur noue delle dette dieci mila parti. E che altro è questo, saluo
che cadendo vna palla di piombo da vna torre alta dugento braccia
trouar, che ella anticiperà vna d' Ebano di manco di quattro dita?
Pesa l' Ebano mille volte più dell' aria, mà quella vescica così gon-
fia pesa solamente quattro volte tanto; l' aria dunque dalla intrin-
seca e naturale velocità dell' Ebano detrae de mille gradi vno, mà
à quella,

Margin annotations:
In aere.
Pl. 10000 — 1.
Eb. 10000 — 10.
Eb. 1000 — 1
vs. 4 — 1

Fig. 4.3. Galileo's discussion of bodies falling in different media,
annotated by Ward. *Savile Bb.13, 76–77. Courtesy of The Bodleian
Libraries, The University of Oxford.*

falling objects with that of air, noting that ebony weighed 1,000 times as
much as air, while the inflated bladder weighed only four times as much.
Ward made note of these values in the lower lefthand margin. In the upper
right margin of the facing page (p. 77), Ward recorded the suppositions of
Galileo's third calculation, comparing the relative velocities of lead and ivory
as they fell through water. In his final annotation in this section, Ward followed Galileo in his calculation comparing the velocity of the same movable

DEL GALILEO. 77

à quella, che pur della vescica assolutamente sarebbe stata l'istessa, l'aria ne toglie delle quattro parti vna: allora dunque che la palla d' Ebano cadendo dalla torre giugnerà in terra, la vescica ne haue- rà passati i trè quarti solamente. Il piombo è più graue dell' acqua dodici volte, mà l'auorio il doppio solamente: l'acqua dunque alle assolute velocità loro, che sarebbero eguali, toglie al piombo la duo- decima parte, mà all' auorio la metà: nell' acqua adunque quando il piombo harà sceso vndici braccia, l'auorio ne harà scese sei. E discor- rendo con tal' regola credo che troueremo l'esperienze molto più aggiustatamente risponder à cotal computo, che à quello d' Aristo- tele. Con simil' progresso troueremo la proporzione tra le velocità del medesimo Mobile in diuersi mezzi fluidi, paragonando non le diuerse resistenze de i mezzi, mà considerando gli eccessi di grauità del Mobile sopra le grauità de i mezzi; ver. gr. lo stagno è mille volte più graue dell' aria, è dieci più dell' acqua; adunque diuisa la velocità assoluta dello stagno in mille gradi, nell' aria, che glie ne detrae la millesima parte, si mouerà con gradi nouecento nonanta noue, mà nell' acqua con nouecento solamente, essendo che l'acqua gli detrae solo la decima parte della sua grauità, e l'aria la millesi- ma. Posto vn solido poco più graue dell' acqua, qual sarebbe, v.gr. il legno di rouere, vna palla del quale pesando, diremo, mille dram- me, altrettanta acqua ne pesasse noue-cen-cinquanta, mà tanta aria ne pesasse due, è manifesto, che posto che la velocità sua assolu- ta fusse di mille gradi, in aria resterebbe di noue cen nouant' otto, mà in acqua solamente cinquanta, atteso che l'acqua de i mille gra- di di grauità glie ne toglie noue-cen-cinquanta, e glie ne lascia sola- mente cinquanta; tal solido dunque si mouerebbe quasi venti vol- te più velocemente in aria che in acqua: si come l'eccesso della gra- nità sua sopra quella dell' acqua è la vigesima parte della sua pro- pria. E qui voglio che consideriamo che non potendo muouersi in giù nell' acqua se non materie più graui in spezie di lei; e per con- seguenza per molte centinaia di volte più graui dell' aria, nel ri- cercare qual sia la proporzione delle velocità loro in aria, e in

K 3 acqua,

in different media, noting that the movable was 1,000 times as heavy as air (thus a ratio of 1,000:1) and 10 times as heavy as water, which translates (in the middle line) to 1,000:100. Ward performed additional calculations on a note slip that he inserted into his 1638 edition and that Wren copied on his own loose sheet (fig. 4.4). On this slip, Ward followed along with the steps of Galileo's first calculation of the fall of lead and ebony in air. He then sum- marized Galileo's calculation of the fall of ivory and lead in water to con- clude, with Galileo, that when lead arrives at the bottom, ivory will only have passed through half its total fall.[38]

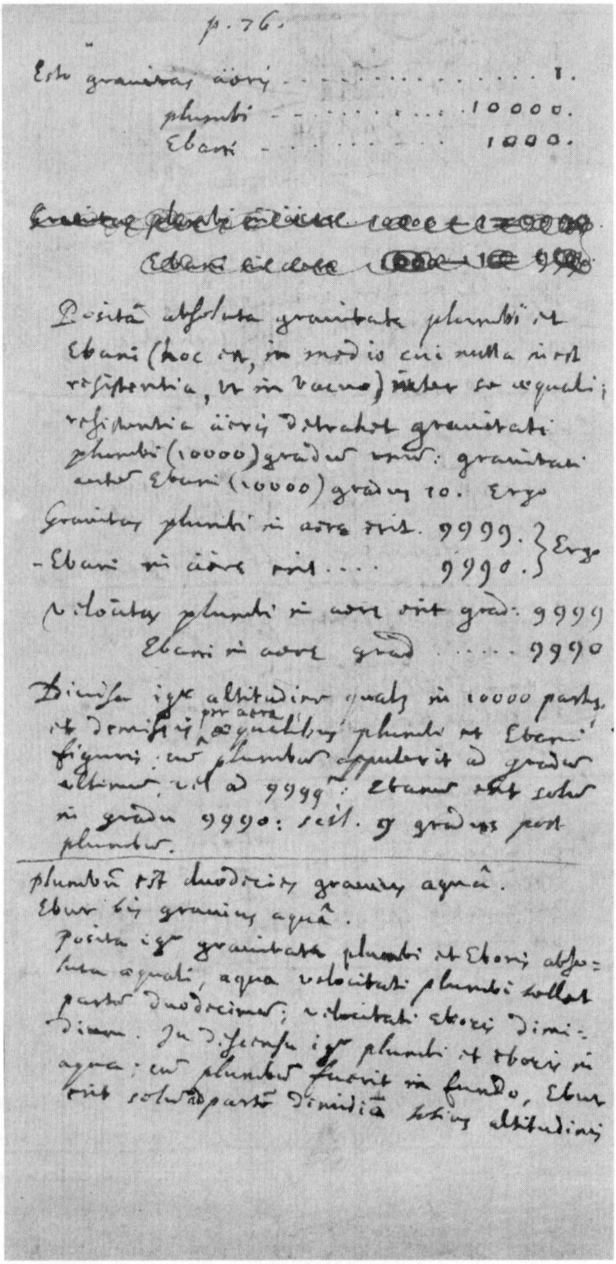
p. 76.

Esto gravitas aëris .... 1.
plumbi .... 10000.
Ebani .... 1000.

Posita absoluta gravitate plumbi et Ebani (hoc est, in medio cui nulla sit resistentia, ut in vacuo) inter se aequali; resistentia aëris detrahet gravitati plumbi (10000) gradum unum: gravitati autem Ebani (10000) gradus 10. Ergo

Gravitas plumbi in aëre erit. 9999. } Ergo
Ebani in aëre erit ... 9990. }

velocitas plumbi in aëre erit grad: 9999
Ebani in aëre grad 9990

Divisa igitur altitudine quali in 10000 partes, et demissis per aera aequalibus plumbi et Ebani figuris: cum plumbum appulerit ad gradum ultimum, vel ad 9999um: Ebanum erit solum in gradu 9990: scil. 9 gradibus post plumbum.

plumbum est duodecies gravius aquâ.
Ebur bis gravius aquâ.
Posita igitur gravitate plumbi et Ebori absoluta aequali, aqua velocitati plumbi tollet partem duodecimam; velocitati ebori dimidiam. In descensu igitur plumbi et ebori in aqua; cum plumbum fuerit in fundo, Ebur erit solum parte dimidiâ totius altitudinis

Fig. 4.4. Note slip showing Ward's scratch work corresponding to Galileo's calculations of bodies falling in different media. *Savile Bb.13, verso of sheet between pp. 18 and 19. Courtesy of The Bodleian Libraries, The University of Oxford.*

In the fall of 1664 the Royal Society carried out a series of experiments dedicated to the subject that so occupied Ward in his annotations of the *Two New Sciences*, namely, the relative velocities of bodies falling in water. These experiments began modestly with the attempt to measure the velocity of small shot as it fell through a narrow tube of 81 inches. On 7 September 1664 further trials were carried out in the same 81-inch vessel, this time to determine the velocity of balls of wax of several sizes as they fell to the bottom. It was found that the smaller balls descended with equal velocity, but the larger ones descended with unequal velocity.[39] Two weeks later, a special apparatus was used that had been constructed for further experiments; it consisted of a square wooden vessel about nine feet high with glass panes on the sides. The descents of balls of wax of three different sizes were timed and compared. Further experiments were carried out by joining a ball of wood to leaden balls of different sizes via a springing wire; the heavier the leaden ball to which it was attached, the faster the wooden ball fell.[40] Galileo had described the fall of such linked objects in the same section of the *Two New Sciences* annotated by Ward; his description was intended to persuade Simplicio that bodies of the same material but different weights will fall at the same speed.[41]

Though the large glass vessel was damaged the next week—one of the glass panes near the top got broken, shortening it to about 7.5 feet—members of the Royal Society continued their experiments by timing and comparing the descent of objects of different shapes, all joined to a leaden weight of two ounces. These trials continued in October for various items, each joined to four ounces of lead. On 12 October the experiments were modified further by varying the weight of a body of the same shape and size; a glass ball was filled with varying proportions of water and shot to compare its time of descent at different weights.[42]

For Ward, annotating these sections of the *Two New Sciences* did not translate into carrying out experiments designed to verify Galileo's claims. Rather, Ward annotated these passages with the goal of accumulating a battery of potential experiments to test. It is worth pointing out that at least occasionally the Royal Society cited their reading of Galileo and others as impulses for conducting experiments.[43] In his letter to Isham and in these annotations, however, Ward does not seem to be describing the reading and subsequent replication of experiments, but a different activity, reading and annotating in a Baconian-inspired fashion for the purpose of culling ideas for potential experiments.

A Mathematical Exercise Book

Whereas Galileo's textual descriptions of experimental procedures prompted this Baconian-inspired collecting of experimental excerpts, the more mathematical sections of the *Two New Sciences* engendered a very different approach. Like pseudo-Baliani and Viviani, Ward and Wren regarded the more mathematical sections of the *Two New Sciences* as a set of mathematical definitions, propositions, and proofs to be worked through. Their focus was less on the content of Galileo's findings and their correspondence with natural phenomena. Instead they read the text as a mathematical exercise book, as a series of mathematical problems to solve, with seemingly little concern about their physical validity. Here Ward and Wren drew on the same techniques of annotation common to sixteenth-century mathematicians as did pseudo-Baliani and Viviani. They followed along with Galileo's proofs using textual and diagrammatic annotations, and they also created alternative demonstrations to and extended Galileo's mathematical claims. One feature that distinguishes the approach of both Ward and Wren from pseudo-Baliani's and Viviani's, however, is their insistence on expressing mathematical relationships in shorthand symbolic form and, in some instances, rewriting Galileo's geometrical proofs using algebraic equations.

In many instances, Ward (and later Wren) followed along with Galileo's text by rewriting Galileo's key assumptions and calculations in a shorthand notation. One example is found at the beginning of Day 2 (fig. 4.5), where Galileo considers the resistance of bodies to breaking and argues that such problems are reducible to the law of the lever. Here a heavy weight is supported partly by the horizontal surface at B and partly by the lever with fulcrum N. Galileo asked readers to determine how much of the total weight of the body is sustained on the horizontal plane and how much by the force at the end of the lever.

Ward annotated this problem with the small diagram at the bottom of the page. The diagram relies on a shorthand notation widely employed in England at the time. Used for expressing fractional relationships, the notation had been popularized by earlier sixteenth-century English mathematicians, including Robert Recorde (1510–1558) and Thomas Harriot (ca. 1560–1621).[44] With this notation, the ratio a:b is as c:d is written as shown in figure 4.6. Galileo proved his assertion by manipulating a number of proportional relationships based on the law of the lever. He began by asking his readers to find a line X such that FB to BO is in the same ratio as NC to X. Ward expressed

DEL GALILEO. 113

Sagr. *Reſto appagato, mà mi naſce vn' altro deſiderio;che è che per intera cognizione mi fuſſe dimoſtrato il modo,ſe vi è, di poter' inveſtigare qual parte ſia del peſo totale quella, che vien ſoſtenuta dal ſoggetto piano, e quale quella, che graua ſu 'l Vette nell' eſtremità A.*

Salu. *Perche poſſo con poche parole dargli ſodisfazzione, non voglio laſciar di ſervirla; però facendone vn poco di figura, intenda V. S. il peſo, il cui centro di grauità ſia A appoggiato ſopra l' Ori-*

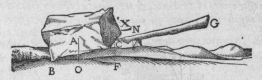

Zonte co'l termine B, *e nell' altro ſia ſoſtenuto col Vette* C G, *ſopra 'l ſoſtegno* N *da vna potenza poſta in* G *; e dal centro* A, *e dal termine* C *caſchino perpendicolari all' Orizonte* A O, C F. *Dico il momento di tutto il peſo al momento della potenza in* G *hauer la proporzion compoſta della diſtanza* G N *alla diſtanza* N C, *e della* F B *alla* B O. *Facciaſi come la linea* F B *alla* B O, *coſi la* N C *alla* X, *& eſſendo tutto il peſo* A *ſoſtenuto dalle due potenze poſte in* B, *e* C, *la potenza* B *alla* C, *è come la diſtanza* F O *alla* O B, *e componendo, le due potenze* B C *inſieme, cioè, il total momento di tutto 'l Peſo* A *alla potenza in* C, *è come la linea* F B *alla* B O, *cioè come la* N C *alla* X, *mà il momento della potenza in* C *al momento della potenza in* G, *è come la diſtanza* G N *alla* N C, *adunque per la perturbata il total peſo* A *al momento della potenza in* G, *è come la* G N *alla* X, *mà la proporzione di* G N *ad* X, *è compoſta della proporzione* G N *ad* N C, *e di quella di* N C *ad* X, *cioè, di* F B à B O, *adunque il peſo* A *alla potenza che lo ſoſtiene in* G, *hà la proporzione compoſta della* G N *ad* N C, *e di quella di* F B à B O, *ch' è quello che ſi doueua dimoſtrare. Or tornando al noſtro primo propoſito, inteſe tutte le coſe*

P   *ſin*

A = c + B   ⎰ C··· G
    FB ⎰ BO
ꞅN··· NC ⎰ X

Fig. 4.5. Problem in Day 2 annotated by Ward. *Savile Bb.13, p. 113. Courtesy of The Bodleian Libraries, The University of Oxford.*

this relationship as shown in Figure 4.7a. As Galileo continued with his proof, stating and manipulating proportional relationships, Ward added them to his diagram. Galileo next claimed that the power B to C is as FO to OB and that A (which is B and C together) to C is as FB to BO. Ward left out the first of these

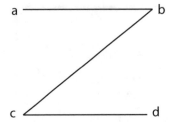

Fig. 4.6. Shorthand notation to express the proportional relationship a:b::c:d.

proportions and added only the second to the diagram (fig. 4.7b). But, Galileo stated, C to G is as GN to NC (fig. 4.7c). Ward superimposed this figure on the developing diagram by adding the additional quantities and connecting them with dotted lines so that the diagram became that shown in figure 4.7d. Because Galileo's goal was to determine the ratio of A to G, Ward underlined A and G and signaled each quantity with a single or double tick mark (fig. 4.7e). A to G, according to Galileo, is in the same ratio as GN to X, so GN and X are also underlined. But by manipulation of the ratios, X is part of a proportional relationship with FB, BO, and NC, so the problem was solved, for Galileo had declared that the moment or force at A to that of G will be equal to the ratio of GN to NC compounded with the ratio of FB to BO.

In rewriting and following along with Galileo's work in the margins, Ward condensed Galileo's dense, prose-filled proofs into a shortened, symbolic form. Aside from the diagrams accompanying his proofs and problems, Galileo's presentation is entirely textual. He wrote and solved his problems in prose using the tools of Euclidean geometry. His mathematical relationships were expressed in words, not symbolic equations or shorthand notation.

Here we can contrast Ward's interest in condensing Galileo's proof into symbolic form with the approach of Viviani and pseudo-Baliani explored in previous chapters. Pseudo-Baliani, for example, annotated this same passage, but he did so by offering a prose summary that followed along with Galileo's text (see fig. 1.2). This prose-rich style of annotation was also used by Viviani as he followed along with Galileo's proofs.[45] Another activity that occupied Viviani and pseudo-Baliani as they read Galileo's mathematics was indicating in the margins how the steps of one proof relied on earlier results. Ward and Wren, in contrast, rarely noted the relationship between Galileo's demonstrations, and they avoided summarizing the steps of Galileo's proof in words. Local training no doubt contributed to these differing modes of annotation. Galileo, for example, wrote the *Two New Sciences* in the

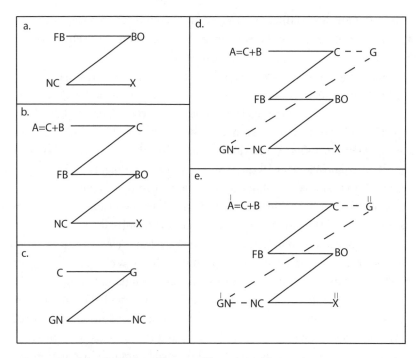

Fig. 4.7. Working through Galileo's problem alongside Ward.

same style as Viviani and pseudo-Baliani annotated it, relying on a prose-rich approach to demonstrations characteristic of Euclidean geometry. We cannot forget, either, that Galileo taught Viviani some mathematics and read the *Two New Sciences* with him.

Other evidence, however, suggests that different methods of mathematics and their related practices were most likely responsible for these variant responses to Galileo's geometrical demonstrations. The prose-rich style used by Galileo and his Italian readers was associated with work in Euclidean geometry, while the symbolic annotation style was employed by readers immersed in the developing algebraic tradition. That these varied styles depended less on geographical location and more on practices conditioned by the aims of varied mathematical traditions is indicated by the annotations of Savile, which survive in numerous books in the Savile Library at the Bodleian. Savile annotated his books, ranging from Euclid to Apollonius, in a manner similar to that of Viviani and pseudo-Baliani, summarizing proofs in the margins and paying close attention to the relationship between steps in individual proofs and earlier results.[46] What distinguishes Savile, Viviani, pseudo-Baliani, and Galileo from Ward and Wren is less geography than the latter's commitment to and involvement in the developing tradition of algebra, as opposed to geometrical demonstration.

Another indication that Ward (and Wren) read Galileo with an eye toward this tradition is Ward's decision to rewrite a handful of Galileo's problems and proofs in a style he denoted by the term ἀναλυτιχῶς (analytically). (Wren copied Ward's work.) This reference alludes to the "analytic art," a term employed by François Viète (1540–1603) to describe his new system of symbolic algebra. Ward provided analytical solutions to four problems in the *Two New Sciences*.[47] In them, Ward followed the basic procedure of Viète's analysis: he supposed Galileo's problem solved, devised an equation following this assumption, and then manipulated the equation to solve it.

Ward's solutions reveal that he used the *Two New Sciences* as a workbook of sorts for practicing or at least working with the techniques of algebra. Viète's methods were introduced to students in England through William Oughtred's (1573–1660) *Clavis Mathematicae*, first published in 1631 and eventually issued in four more Latin and two English editions. Oughtred's *Clavis* was well known to Ward and Wren, and its methods were promoted in Oxford mathematical teaching. Ward received personal instruction from Oughtred as a fellow of Sidney Sussex College, Cambridge, in the early 1640s, and he was instrumental in seeing the second and third editions of the *Clavis* into print for use as a textbook at Oxford. Wren also belonged to this community of English mathematicians interested in promoting analysis. Oughtred mentioned him in the third edition of his *Clavis*, to which he also appended a translation made by Wren of a treatise on sundials.[48] Ward's and Wren's explicit references to Viète in their annotations signal their commitment to the analytic art, or new algebra, in the style of Viète and Oughtred. Their pursuit of analysis in this tradition was shared by other Savilian professors, including Wallis, who held the chair in geometry from 1649 to his death.[49]

Ward's solutions also confirm his interest in the mathematical aspects of Galileo's findings, not their physical or philosophical implications. Ward, for example, tended to abstract from Galileo's solutions the mathematical relationships central to the problem; their physical implications went unmentioned. To consider one instance, in Proposition 14 of Day 3, Galileo asks readers to find the part (AX) of a vertical line (DB) that is traversed by a movable from rest in the same time in which the same movable traverses an inclined plane (AC) after traveling through the same vertical (fig. 4.8). While

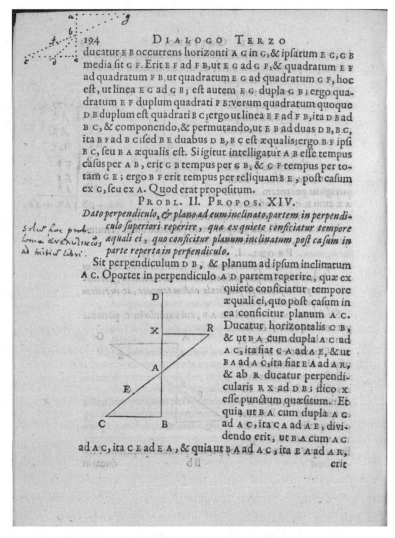

Fig. 4.8. Image accompanying Problem 2, Proposition 14, in Day 3, annotated by Ward. *Savile Bb.13, 194. Courtesy of The Bodleian Libraries, The University of Oxford.*

part of Galileo's solution involved manipulating geometrical relationships, much of his proof was devoted to comparing times and distances of falls using Galileo's quantitative rules for naturally accelerated motion. In his analytical solution, in contrast, Ward focused entirely on the proportional relationships between the line segments. He ignored nearly half of Galileo's proof, leaving unmentioned Galileo's demonstration regarding the paths traversed by heavy bodies in natural motion.

Ward's appropriation of Galileo's *Two New Sciences* as a text for practicing the analytic art calls to mind Isaac Casaubon's use of Bacon's *Advancement of Learning* as a means of practicing his English. Casaubon, like Ward and Wren with the *Two New Sciences*, annotated Bacon's text with standard philological techniques, paying attention to Bacon's rhetorical devices and parallel passages from other works. However, Casaubon also marked (or had someone mark for him) the accented syllable of every syllabic word, a practice that William Sherman concludes indicates that Casaubon may have read the text aloud, in part to practice his pronunciation of English.[50]

Casaubon and Ward found uses for the books of their near contemporaries that are not immediately obvious to the modern historian of science, who often comes to these landmark texts focused more on their arguments and novel advances than on what they offered for immediate practical use. Ward's interest in Galileo's mathematics to the exclusion of his physical arguments, moreover, reveals a similar intention as does his decision to work through Galileo's problems in more symbolic language. Both features of Ward's mathematical reading suggest that he regarded the quantitative parts of the *Two New Sciences* as a mathematical exercise book.

This practice of reading contemporary geometrical texts and rewriting them using analytical techniques bears a strong resemblance to later readers' reactions to Isaac Newton's 1687 *Principia*. The Scottish mathematician David Gregory (1659–1708), for example, reworked many of Newton's geometrical demonstrations with symbolic, analytical tools, rewriting several of Newton's proofs in terms of the analytical method of fluxions. In other cases, he translated Newton's prose into algebraic terms.[51] Gregory's rewriting, in turn, is only one among many such applications of analysis to Newton's geometrical methods. The results of such rewritings at the hands of mathematicians, including Pierre Varignon (1654–1722), helped to disseminate Newton's ideas in the eighteenth century.[52] Historians of science have often considered these translation projects in the context of an eighteenth-century commitment to analysis or a quest to develop a geometrical picture of the physical world that turned to analysis in order to deal with the problem of the infinite.[53] While these characterizations are no doubt correct, I want to emphasize the scholarly practices underlying these ideological positions. Choosing to translate Newton's *Principia* into the language of calculus involved a particular type of active reading by individuals committed to the developing analytical algebra and calculus. Ward's and Wren's rewritings of Galileo's problems in their annotations suggest that this reading practice may have been widespread among advocates of the new style of mathematics. Further research may reveal that these later translations of Newton's readers were inspired by earlier methods of mathematical reading that encouraged this type of translation activity as a means to practice and perfect one's skills in analysis.

A Work of Traditional, Textual Natural Philosophy

Just as the discipline-specific practices of mathematical reading shaped the way Ward and Wren worked through the mathematical portions of the *Two New Sciences*, so, too, did other scholarly practices, particularly those associated with more textual, bookish disciplines, help them make sense of the book's more descriptive sections. Many of the annotations that Ward recorded and that subsequently were copied by Wren, for example, are topical notes identifying subjects central to traditional natural philosophy as inherited from the medieval university. Ward noted Galileo's discussion of such topics as "the resistance of the void," "on infinite indivisibles," and "examples of rarefaction and condensation."[54] He also recorded Galileo's conclusions regarding these topics, including the opinion that the continuum is made up of indivisible atoms, that the velocity of a descending body does not depend on its heaviness, that air is heavy, and that a heavy body in motion differs from one at rest.[55] All of these were standard topics discussed by Aristotle and addressed in university teaching through the seventeenth century and beyond.[56]

This practice of noting general topical headings and the main conclusions advanced in a text was one common to scholarly reading in multiple genres in the medieval and early modern periods. Surviving copies of Herodotus, Pliny, and Terrence annotated by Casaubon contain marginalia with topical headings and other notes related to the texts' language and rhetoric.[57] Some readers apparently read Galileo's *Dialogue* in a similar way, noting in the margins paraphrases of Galileo's argument.[58] Ann Blair has argued that this flagging of passages of interest for the purposes of retrieval and retention accounted for the bulk of reading notes left in the margins of printed books in the early modern period.[59]

The attention Ward paid to Galileo's discussion of standard philosophical topics is also revealed by the longer notes he included in the inside covers of his book. Wren too was interested in this aspect of Ward's reading, for he copied these longer sections on note slips that he inserted into his copy. Ward, for example, included on the inside cover a more extensive summary of Galileo's explanation of rarefaction and condensation under the title "Condensation and Rarefaction according to the opinion of Galileo." In this note, Ward summarized Galileo's claim that the two processes could be explained by supposing that bodies were composed of infinite indivisibles.[60]

Ward's annotations also reveal his concern with a central theme of traditional natural philosophy, namely, a search for the causes of natural phenomena. In Day 3 Ward included the marginal note, "The cause of accelerated motion in the descent of heavy bodies."[61] In the corresponding passage, Galileo's interlocutor Sagredo briefly mentions some common speculations on the cause of accelerated motion. Salviati quickly interrupts, declaring that such a topic lies outside the Academician's intended aims to describe motion quantitatively. Ward's annotation suggests that he read this passage as Galileo's summary of possible causes of accelerated motion, not as a statement of Galileo's intent not to treat this subject.

An additional annotation in Day 4 confirms that Ward read the *Two New Sciences* with this question of causes in mind, even though Galileo had explicitly advised readers to reject just such an approach.[62] In the relevant passage, Sagredo compares Galileo's conception of the horizontal and vertical components of the projectile's parabolic trajectory to Plato's idea that God had started all planets moving toward the Sun from the same point in the universe and later converted their rectilinear motion into a uniform circular one. In the margin of his copy, Ward noted that if the Platonic hypothesis to which Galileo refers was correct, it would be necessary "that the Earth and other planets pay heed to some heavenly body (perhaps the Sun) as a center in their proper motion (just as our heavy bodies are carried to the Earth as a center) or by a different route some certain thing must be assigned as a cause of accelerated motion."[63] Whatever Ward's thoughts were regarding the possible causes of accelerated motion and of planetary motions, his annotation reveals that contrary to Galileo's own professed intentions, he was interested in the causes underlying projectile, accelerated, and planetary motion and that he brought this concern with him as he read the *Two New Sciences*.

In addition to their consideration of the standard topics and terms of traditional natural philosophy, these readers were attentive to the relationship between Galileo and other textual authorities. Many of Ward's annotations reference the works of other authors, including Christoph Clavius and Bonaventura Cavalieri, as well as Galileo's own previous publications.[64] At

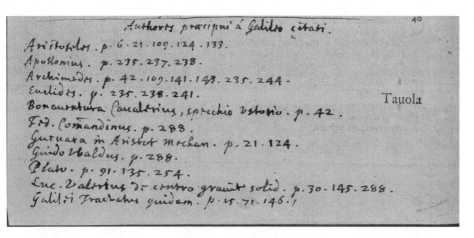

Fig. 4.9. Ward's list of authors cited by Galileo. *Savile Bb.13, table of printing errors (detail). Courtesy of The Bodleian Libraries, The University of Oxford.*

times Ward signaled Galileo's relationship to Aristotle. Alongside the section of Day 1 in which Galileo criticizes Aristotle's opinions on the relative velocities of falling bodies, for example, Ward included a marginal annotation that reads, "Aristotle's error."[65] Wren included a separate annotation, not found in Ward's text, that reads, "this is characteristic of Aristotle," alongside a passage in which Galileo explains that the coherence of bodies can be explained by nature's abhorrence of the void acting between the smallest particles of bodies.[66] Ward confirmed his interest in Galileo's relationship to other textual authorities by including a "list of authors cited by Galileo" in his copy; Wren transcribed the list into his own book (fig. 4.9). This list includes Aristotle, Guidobaldo del Monte, Euclid, Plato, and Galileo himself.

Attention to authorities was a common feature of note-taking in the sixteenth century and earlier. Copernicus's readers often used their marginalia to signal the sources of Copernicus's arguments, in addition to flagging related passages of interest. Erasmus Reinhold, for example, annotated his copy of Copernicus by noting the author's sources, including Giorgio Valla (1447–1500), Johannes Werner (1468–1522), and Regiomontanus (1436–1476), even when Copernicus himself did not identify them explicitly. Paul Wittich annotated some of his copies with cross references to other passages of *De Revolutionibus* and included references to other authors' discussions of parallax.[67] In the editions of Benedetti and Jordanus of Nemore that he annotated, the mathematician Guidobaldo dal Monte also underlined and made note of textual authorities, including Aristotle and Archimedes, who were cited in the text.[68]

Creating a separate list of authors cited, a "list of authorities," or *catalogus auctorum*, as Ward and Wren did, seems to have been a specialized note-taking practice more specific to textual genres than to mathematical ones, at least given available evidence. Copernicus's readers, though they noted textual sources in the margins, did not compile such lists. Rather, the *catalogus auctorum* was a common paratext accompanying compilations, especially legal ones. Justinian had mandated in the sixth century that the *Digest* include a list of the authors from whom the extracts were made in order to legitimate the compilation made from them. Medieval manuscripts of the *Digest*, as well as other compilations, generally included such a list, either in the margins of the manuscript or in the text itself. Early modern printed compilations, including Theodor Zwinger's (1533–1588) *Theatrum vitae humanae* (first published in 1565) and various editions of Desiderius Erasmus's (1466–1536) *Adages*, all included lists of authors cited. These lists could consist only of names as a testament to the quality of the work, or they could comprise page numbers or other information designed to be useful to the reader.[69] Some of the readers of Bodin's *Theatrum* also included lists of authors cited. Casaubon, for example, recorded a select catalogue of Bodin's criticisms of Aristotle on one of the three flyleaves he used for notes; he also kept a select record of other authorities mentioned by Bodin.[70] In his 1577 edition of Isidore of Seville's *Etymologiae,* Bonaventura Vulcanius (1538–1614) compiled such a *catalogus auctorum* as a testament to his own readers of the authority of his production, one grounded in Isidore's appropriation of the textual authority of earlier authors.[71]

That Ward and Wren each created such a list thus indicates their application of note-taking tools specific to textual and descriptive genres, such as law and natural philosophy, to the *Two New Sciences*. It also suggests their continuing adherence to forms of scholarship that valued the authority granted to texts based on the process of textual transmission and citation. It has been argued that these lists were not merely finding aids, intended to facilitate readers' navigation of the text in question. Rather, they served the purposes of citation and demonstration of authority, showcasing to readers the validity of a work based on the lineage of textual claims in which its author positioned it.[72]

In general, Ward and Wren applied such reading strategies associated with textual genres primarily to Day 1 of the *Two New Sciences*, and they used the more mathematical approaches in Days 2 through 4. This uneven attention to different parts of the *Two New Sciences* parallels the reaction of sixteenth-century readers of Copernicus. Reinhold, for example, left the cosmological sections of *De Revolutionibus* unglossed but made extensive annotations to the more technical sections dealing with precession, the relative motion of the Earth, and chronology.[73] Ward and Wren, in contrast to Reinhold, annotated almost the entirety of the *Two New Sciences*, but their varied annotation styles reflect their judgment that the book drew on different disciplinary traditions and required corresponding and distinct styles of reading and note-taking.

A Textbook

When Ward turned to the *Two New Sciences*, he did so with a variety of goals in mind, applying a diverse set of reading methods to the text. Wren no doubt imbibed many of these aims and methods as he copied Ward's annotations, but the context in which he did so differed. Wren's references to Ward's demonstrations resemble those of a pupil copying the lessons of his master.[74] The means of transmission of this course would have involved the copying of marginal annotations. Thus we have the last and perhaps most obvious manner in which the *Two New Sciences* was read by Ward and Wren: as a pedagogical text.

Glossing, summarizing, commenting, and the copying of such explications—via student lecture notes or through marginalia—were activities central to both Scholastic and humanist educational practice.[75] The summaries of Galileo's arguments that Ward offered in the margins and flyleaves, especially for the material in Day 1, conform to a more general pattern of teaching by paraphrasing or epitomizing, especially in literary genres. In the words of the humanist Latin teacher Antonio della Paglia (ca. 1500–1570), paraphrasing involves repeating the passage "with other words, sometimes less, sometimes more, sometimes an equal number [in order to] convey gracefully the same meaning as the author."[76] In his study of a set of pamphlet editions of classical texts, Anthony Grafton reconstructed how the lecturer Claude Mignault (ca. 1536–1606) used such techniques to teach the classics at the University of Paris in the early 1570s. According to the student notes festooning the margins of these extant pamphlets, Mignault provided a word-for-word paraphrase in Latin, which several students recorded in a tiny script between the lines of the printed text.[77]

While this style of teaching is often associated with the texts of classical authors, the use of a technical mathematical work in early modern pedagogy is not unknown. Many of Copernicus's readers acted similarly, using marginalia and copying marginalia from one copy to another, as they responded to the first two editions of his *De Revolutionibus*. As with Wren's copying of Ward's analytical solutions, at times Copernicus's readers glossed and then transcribed annotations that reflected their own mathematical models, devised in conjunction with the original text but not based on it.[78] Many of these annotations, such as those of Wittich, were likely intended as instruments of private study as well as instructional tools. Like Wren's copying from Ward, examples exist in which students or colleagues copied out key annotations from other exemplars of *De Revolutionibus*.[79] The Swiss humanist Heinrich Glarean used his own publications on musical theory as the basis for lectures he gave at the University of Freiburg toward the end of his career. Surviving student notes reveal that these students used catchwords pointing to the key points in the text, cited the sources of the content of particular passages, and often included additional quotations from these sources.[80] Similarly, students at the University of Paris who commented on Sacrobosco's *Sphere* mined its astronomical content to aid in their explications of other literary texts and critically compared authorities on astronomy.[81]

While other Savilian professors whose books were donated to the professors' library, including Savile and Wallis, are known to have been avid note-takers, Wren does not appear to have been a compulsive annotator.[82] Invariably, books donated by Wren contain no marginalia.[83] Even the 1655–56 *Opere*, which contains Wren's annotations of the *Two New Sciences*, is clean, apart from the annotations derived from Ward. His extensively annotated *Two New Sciences* is thus an exception. Its atypical nature further suggests that its marginalia were not the product of Wren's everyday scholarly practice but were occasioned by an outside influence, such as a pedagogical setting.

By copying his annotations, Wren read the *Two New Sciences* through Ward's eyes. He made note of the aspects of the text Ward thought important to annotate, and he followed along with Ward as he worked through Galileo's mathematical proofs. In addition to absorbing Ward's interpretation of the *Two New Sciences*, by copying Ward's annotations Wren simultaneously received instruction in how to read. Ward read the *Two New Sciences* in multiple ways and used the text as a springboard for a variety of disparate activities, from mathematical calculations to the development of new experimental techniques. Some of these methods of reading, such as note-taking on standard natural philosophical topics, were widely practiced and easily learned from other individuals. However, other aspects of Ward's reading, notably his appropriation of the humanist tools of note-taking and commonplacing for the purposes of experimenting, seem to have been novel and, at least judging by Ward's 1652 letter to Isham, in need of some description and explanation. By copying Ward's annotations, Wren would have absorbed these new reading methods firsthand and seen how to apply them in his own work.

Why Annotate the *Two New Sciences*?

Historians have often proclaimed seventeenth-century Englishmen's inherent sympathy for Galileo. According to these accounts, Galileo's publications were received with an enthusiasm that was fueled both by his growing fame and by his difficulties with the Catholic Church. Following the printing of the *Starry Messenger*, Galileo's astronomical writings became a standard point of reference for English scholars, and the telescope became an indispensable tool for astronomical observations. By the 1620s a host of prominent English mathematicians and savants—including Henry Briggs (1561–1630), Mark Ridley (1560–1624), John Bainbridge (1582–1643), and Harriot—were carrying out their own telescopic observations. Galileo's condemnation in 1633 disposed many Englishmen even more favorably toward him, for it made Galileo, in their eyes, a symbol of opposition to the Catholic Church. In 1634, for example, Thomas Hobbes (1588–1679) remarked that in Italy Galileo's *Dialogue* was considered a book that would do more harm to religion than had all the books of Luther. John Milton (1608–1674) cited Galileo's fate as an example of oppression in his 1644 *Areopagitica*, while the reverend Jeremy Taylor (1613–1667) noted that the Catholic Church's judgment against Galileo would make it a point of ridicule for future generations.[84]

While Ward's and Wren's abundant annotations of the *Two New Sciences* are in keeping with this accepted narrative of an English enthusiasm for the Pisan philosopher and mathematician, there is little indication that Galileo's astronomical work shaped their reading of the *Two New Sciences*. Ward's and Wren's marginalia contain no reference to Galileo's earlier astronomical publications, including relevant passages of the *Dialogue*. The only allusion to cosmological matters at all was in response to Galileo's reference to Plato in his discussion of projectile motion. Ward and Wren read the *Two New Sciences* with enthusiasm, but they apparently did so without considering the cosmological implications of the *Dialogue*. In fact, they read the work largely without taking into account Galileo's physical arguments in the *Two New Sciences* at all.

The abundant annotations in the copies read by Ward, Wren, Viviani, Mersenne, and pseudo-Baliani are outliers in the context of surviving readership records of Galileo's publications. The vast majority of copies of the *Two New Sciences* I have identified were annotated sparingly, if at all. Richard Allestree (1619–1681), Regius Professor of Divinity at Oxford from 1663, for example, owned a copy of the *Two New Sciences*, which he bequeathed, along with the rest of his library of around three thousand volumes in 1681, to Christ Church, Oxford.[85] While many of the books in the Allestree Library were annotated, either by Allestree or by other professors or by students of theology, his 1638 *Two New Sciences* is not. The absence of annotations in Allestree's copy is perhaps a reflection of its niche audience, for Allestree's library was intended to aid future professors in the defense of the Church of England. Galileo's book likely contained little of interest to students and professors of theology.

Other seventeenth-century scholars who maintained personal libraries and are known to have been note-takers or annotators also owned copies of the *Two New Sciences* that they did not annotate. August, future Duke of Brunswick, as a boy flagged and underlined passages in his reading and then copied them into separate notebooks.[86] Though he purchased a copy of the 1655–56 *Opere* containing the second printing of the *Two New Sciences*, the absence of annotations in it suggests that he did not read it with the same attention he gave to his schoolbooks.[87] Sir Kenelm Digby (1603–1665), the English courtier and diplomat known for his correspondence with Fermat and Mersenne as well as for his interest in alchemy, accumulated two separate libraries during his lifetime. His copy of the 1638 *Two New Sciences* belonged to the second library, which he assembled while living in Paris.[88] Digby's copy contains no annotations, even though we know that he was familiar with the work—he referred to it in his 1644 *The Nature of Bodies*—and took notes in other contexts, notably in the manuscript collection he bequeathed to the Bodleian Library in 1635.[89] Other seventeenth-century scholars who owned but did not annotate a copy of the *Two New Sciences* include the Swiss diplomat and scholar Ezechiel Spanheim (1629–1710) and St George Ashe (ca. 1658–1710), provost of Trinity College, Dublin, and a member of the Royal Society.[90]

The lack of annotations in surviving exemplars of the *Two New Sciences* mirrors the findings of other scholars in regard to Galileo's earlier publications. Extant copies of his 1610 *Starry Messenger* contain few annotations even though the book was printed with deliberately large margins, ostensibly to give it the appearance of a luxurious product but one that also would have facilitated note-taking.[91] Surveys of extant copies of the *Dialogue*, whose publication and subsequent prohibition were also widely publicized, similarly have revealed few heavily annotated copies.[92]

What, then, prompted these five readers—pseudo-Baliani, Viviani, Mersenne, Ward, and Wren—to annotate the *Two New Sciences* so extensively? Unlike many scholars of the previous century, notably the philologist Casaubon, the courtier and writer Gabriel Harvey, and even the astronomer Reinhold, these readers gave no indication in their annotated copies of their purpose, their method, or the circumstances in which they took notes on Galileo's book.[93] With the exception of Mersenne's cursory reflections in his prefatory material and Ward's letter to Isham, none of these readers even followed the practice of Boyle, who did include scattered comments throughout prefaces, advertisements, works, notebooks, and manuscripts that described how he worked with texts.[94] This silence suggests that these readers approached the reading of and taking notes on the *Two New Sciences* as an implicit and unproblematic aspect of their scholarly practice, one that, having been internalized, required no additional reflection or comment. Such an attitude also may have been prompted by the type of text being read. Harvey's attention to the circumstances and purposes of his own reading no doubt was encouraged by the fact that many of the texts to which he turned his eyes and pen contained commentaries and instructions on the correct methods of reading.[95] While the format of Days 3 and 4 of the *Two New Sciences* can be read as an implicit model for Galileo's readers, Galileo himself offered no programmatic statement informing readers how he hoped his text would be approached.[96]

A number of factors seem to have inspired these readers to annotate the *Two New Sciences*. Ward, Wren, pseudo-Baliani, and to some extent Viviani seem to have been drawn to the text for the subjects it addressed, flagging topical heads related both to traditional natural philosophy and to mathematical subjects. Ward turned to the *Two New Sciences* as part of his activities with the Oxford Philosophical Club. For many of these readers, the *Two New Sciences* served as a depository of annotations because it served a pedagogical function, either as a place to copy the annotations of a mentor or as a place to work through mathematical problems. In Mersenne's case, copies of the *Two New Sciences* stored the treasured observations of his correspondent Descartes. This variety of approaches to the text derives from the diversity of readers' experiences and motives. Mersenne annotated to preserve Descartes's opinions in a useful format. Viviani annotated as part of his reading as an editor and student. Ward came to the text with multiple purposes and annotated it accordingly; Wren then copied as a student. Pseudo-Baliani apparently approached the book with many of the motivations often attributed to readers of Galileo: as a critical peer with a desire to further Galileo's program and interrogate his assumptions.

Apart from pseudo-Baliani's, few of these reactions match the current portrait of the reception of the *Two New Sciences*. Nor do they correspond with Galileo's depictions of his own scholarly methods, which he portrayed as using mathematics and real-world experience to validate textual claims. What emerges from these readers' surviving marginalia is their judgment that the text belonged to multiple genres and merited reading as such. They saw traditional reading methods common to a variety of genres in the sixteenth century as applicable to it. At the same time, they appropriated the text for a variety of purposes, using it as a source for new experiments, as a mathematical exercise book, and as a pedagogical textbook. In this sense, we can see in these examples a type of active, goal-oriented reading similar to that ascribed to other period scholars.[97] Just as early modern readers of history, philosophy, and literature returned to the same texts but for different purposes, seeking to use them for some other outcome beyond the mere accumulation of information, so, too, these readers read and annotated the *Two New Sciences* to further their pedagogical, natural philosophical, experimental, and mathematical pursuits, many of which bore little resemblance to Galileo's own.

CHAPTER FIVE

The University of Pisa and a Dialogue between Old and New

Surviving marginalia give evidence of readers at work as they puzzled through the mathematical demonstrations and textual arguments of Galileo's *Two New Sciences*. The readers discussed in previous chapters approached Galileo's book using a variety of methods, which testifies to the multiple meanings and uses it held in the seventeenth century. One key feature of their reactions was their application of traditional scholarly practices to the *Two New Sciences* and a focus less on the validation and extension of its claims and more on its utility for enterprises based either on textual comparison or on mathematical problem-solving. These findings contrasted with the accepted narrative of the book's reception, which has painted a picture of readers intent on verifying or invalidating its conclusions. Yet while surviving annotations reveal much about how individual readers approached Galileo's book, in most cases they tell us little about the diffusion of readers' interpretations and the larger scholarly community in which the text was read.

This chapter and the next turn to complementary evidence, the surviving notes and teaching materials of university professors, to address precisely these questions of reading methods, dissemination, pedagogy, and intellectual community within an institutional setting. Recovering evidence of reading practice from these teaching texts is a less straightforward task than examining marginalia. I proceed from the premise that the details of these texts are significant. Not only when professors cited Galileo's work and whether they agreed or disagreed with him are important, but in what sections of their curriculum they did so and how they incorporated their citations of Galileo's text into their exposition also matter. Comparisons between editions of the same teaching text, between an author's citations of various contemporary authors, and between teaching texts for different institutions often provide important context for understanding the reading practices that led professors to teach Galileo in the form that they did. These chapters embrace the insight that teaching provides evidence both of how professors read Galileo and of how they modeled the reading of a self-styled innovator like Galileo to their students. They also recognize that not all professors might have read the *Two New Sciences* directly; at times professors relied on intermediate sources in composing their teaching texts. Because this was an important way that period readers learned of Galileo's ideas, I pay special attention to these issues of transmission.

The University of Pisa has been portrayed in Janus-like fashion, alternatingly as a conservative or an innovative institution. Reflecting the traditional view of early modern universities as sterile, moribund institutions against which novel currents of seventeenth-century science emerged, some historians have depicted the Pisan university in the seventeenth century as one in decline.[1] Their accounts have emphasized the conservative, reactionary nature of the institution by stressing the animosity between supporters of Galileo and university professors at Pisa and elsewhere.[2] In other venues, scholars have portrayed the University of Pisa in a more positive light, emphasizing the openness of its faculty, who embraced Galileo's legacy even in the face of restrictions imposed by the Catholic Church. A significant number of Pisan professors of natural philosophy and mathematics advertised their interest in Galileo's work and other current trends in philosophical thought, especially atomism.[3] Two professors, Carlo Rinaldini (1615–1698) and Pascasio Giannetti (1660–1742), played a role in the publication of the first two editions of Galileo's *Opere*.[4] Pisan professors also interacted with the larger Tuscan intellectual community, including the recently formed Accademia del Cimento, which consciously undertook a program of experimentation inspired by Galileo.[5]

In the spirit of the latter interpretation, Pisa has often been described as an institution that continued Galileo's battles over the correct method and goals of studying the natural world. According to this narrative, innovative professors were so eager to incorporate Galilean and atomistic doctrine in their research and teaching and had such success that they aroused the ire of their more traditional colleagues. The result was a conflict initiated by the more conservative members of the Pisan faculty, led by Andrea Moniglia (1625–1700), Luca Terenzi, and Giovanni Maffei, against followers and admirers of Galileo, most notably Lorenzo Bellini (1643–1704), Donato Rossetti (1633–1686), and Alessandro Marchetti (1633–1714). Though Marchetti and his compatriots defended themselves admirably, according to this narrative, nothing could hold back the anger of the traditionalists. The controversy ended with a Peripatetic victory: in 1670 a de facto prohibition was issued preventing the teaching at Pisa of atomist and Galilean doctrine. This proscription was made more severe in 1691 by a provision enacted by Cosimo III condemning the doctrine of Democritus.[6]

Extant documents do reveal that members of the so-called conservative faction and the opposing innovators offered specific, programmatic statements that asserted fundamental differences between their approaches to the study of nature. These rhetorical statements support the view that the Pisan intellectual community portrayed itself as a battleground between traditionalists and innovators. For example, in one letter, Maffei admitted that Aristotle was known to have made errors in his writings and that it was necessary to moderate sensory experience, which was inherently fallible, with reason. In the midst of his discussion, however, the purportedly conservative Maffei clarified that he regarded the repeated, naturally occurring experiences advocated by Aristotle, not the artificial experiments carried out by his contemporaries, as appropriate to the discipline of natural philosophy.[7] In his words, "When I say experiences, I mean natural ones, that is, those that are seen daily in the world [and] carried out by nature. . . . Effects diverse from natural ones are seen in those experiences that are carried out artificially by men in the way that they like best."[8] In the same letter, Maffei made another pointed remark that could be taken as a sign of a gulf separating his approach to the study of nature from that of his colleagues: "I say, in addition, that experience shows the effect but it doesn't reveal the cause of it; and besides *scienza* consists not in the cognition of the effects but in the knowledge of the causes of [such effects]."[9] With these two statements, Maffei voiced his objection to two characteristics often associated with novel approaches to the study of the natural world in the period: the substitution of traditional, universal experiences by contrived event-experiments and a reevaluation of the proper goals of natural philosophy to place less emphasis on causes and more on their effects and prediction of them.

The so-called innovators were equally skilled in arguing for their position by emphasizing the differences in intellectual outlook and the unreasonableness of their opponents. Defending the position of those labeled as "atomists," Marchetti drew a sharp distinction between the methods he and his supporters embraced and those espoused by his more traditional colleagues. Quoting Aristotle, who had famously said that he was the "friend of Socrates and of Plato, but the best friend of Truth," Marchetti claimed that contrary to the accusations leveled against them, he and his fellow "atomists" studied a wide variety of authors, "the opinions of Democritus, and of Epicurus, and of Anaxagoras, and of Plato, and of Aristotle, and of Galileo, and of Gassendi, and of all other well-known philosophers," but that they did not study them "to follow one more than another, as the Peripatetics do," giving greater weight to the *ipse dixit* of Aristotle than to sensory evidence and demonstrations. He went on to argue that his Peripatetic colleagues "laugh, mock, and tease all the others, and except Aristotle alone, they hold all other philosophers as beasts, not condoning even the divine Plato, the master of their master, while it appears almost the same thing to the Peripatetics to say that such an idea is from Plato as it is [to say] that it is a dream, a chimera, a poppycock, a castle in the air." "Impartial philosophers," in contrast, "with the greatest diligence weigh with the balance of the assayer the opinions of each one." Their enterprise was designed as an "industrious apiary," in which "the good and sweet of the truth and the credible" were drawn from the writings of others. They left behind "the bitterness and poisonous of falsity and improbability to the wasps and hornets," who made sterile and unhappy honeycombs from their sophistries.[10]

In their contributions to the exchange, Marchetti's allies portrayed their opponents' devotion to Aristotle as relentless to the point of absurdity. Bellini described hearing from a flood of people that Terenzi and his supporters "were going from bridge to bridge and piazza to piazza spreading the word that I didn't teach ancient books." Such an assertion, however, did not correspond to reality, for Bellini promised that he "faithfully and with rheum [*flemma*] carried the texts of those authors into the lecture theater and read them publicly."[11] Giovanni Alfonso Borelli (1608–1679) too protested that he and his friends had not incited their adversaries; rather, they were victims, "abused and publicly scorned with so much insolence" that they were required to defend themselves.[12]

Narratives based on these rhetorical pronouncements have portrayed the University of Pisa as an institution in which two camps pursued separate philosophical projects. The type of reading of the *Two New Sciences* observed in the previous chapters, in which readers applied an eclectic set of methods for a variety of intellectual projects, would not be possible at Pisa, according to this depiction. The camp of innovators would wholeheartedly embrace Galileo's novel methods and apply them to his work, while those more bound to traditional approaches would categorically reject the book without engaging with it.

The scholarly practices of these professors, gleaned from evidence of actual reading and teaching, suggest otherwise. Like the annotators discussed in previous chapters, readers at Pisa read the *Two New Sciences* using a variety of methods. At times they applied the book to scholarly projects that seem more novel; in other instances their reading was oriented to support traditional natural philosophy or mathematics. The teaching of three professors, Claude Bérigard, Vincenzio Renieri, and Pascasio Giannetti, illustrates this conclusion and offers a window into their reading practices. Whereas Bérigard and Renieri employed traditional natural philosophical and mathematical methods as they read Galileo's book, Giannetti approached the text with a more varied set of practices derived from the developing genre of physicomathematics. The reading practices underlying professors' continuing adherence to traditional approaches and goals are evident in the extant reading notes of Bellini.

The differences separating traditional and innovative members of the faculty, this chapter argues, lay more in their rhetoric than in their scholarly practices. Professors' responses indicate that like the prolific annotators in previous chapters, they saw the *Two New Sciences* as a work of disparate genres to which separate sets of reading methods were applicable. Their tendency to work in discipline-specific ways and to read topically, moreover, led them, for the most part, to read the *Two New Sciences* apart from the *Dialogue*. The teaching materials of these university professors reveal the eclecticism underlying both their interpretations of Galileo and the reading methods they disseminated to and demonstrated for their students.

Putting Aristotle and Galileo in Dialogue

The surviving texts of Bérigard, who taught natural philosophy from 1627 to 1639, suggest that he was the first Pisan professor to address the *Two New Sciences* in his teaching. He also has claim to the dubious honor of being the first person to respond in print to Galileo's 1632 *Dialogue*; his *Doubts on Galileo's Dialogue* was published in the same year. Bérigard had studied natural philosophy and medicine in Aix before being summoned to the Tuscan court in 1626 to serve as a secretary to Grand Duchess Christina. The following year, at the urging of Christina's confessor, Bérigard was appointed an extraordinary professor of natural philosophy at the University of Pisa. Seven years later, he took over the chair previously held by Scipione Chiaramonti (1565–1652) to become an ordinary professor of natural philosophy, a position he held until 1639, when he moved to the University of Padua. A post at prestigious Padua was clearly a move up for Bérigard, but his desire to leave Pisa may have stemmed in part from more personal reasons related to his critique of the *Dialogue*. His short text—it ran to only seventy pages—was addressed as an open letter to the members of the Accademia dei Lincei, the Roman academy that had welcomed Galileo into its ranks in 1611. Bérigard's motivations for writing are unclear. It has been suggested that he wrote at the instigation of the Medici to diffuse a tense political situation between Tuscany and Rome. Whatever the reason, Bérigard's critique of Galileo permanently soured his relations with the Tuscan intellectual community, many of whose members remained ardent supporters of Galileo. Bérigard's perceived animosity may have contributed to his decision to accept the new position.[13]

After moving to Padua, Bérigard published a course on natural philosophy, *Circulus Pisanus*, in 1643. Connections to Pisa were emphasized through the work's title, its dedicatees, and its contents. The common title to each work, *Pisan Circle of Claudius Berigardus Molinens, Once at the Pisan, now at the Paduan Lyceum*, linked Bérigard to the two Italian universities where he taught natural philosophy. Several of his dedicatory letters also stress the twelve years he taught at Pisa as evidence for his competence to produce such a text.[14] The *Circulus* was printed in four separate volumes, each dedicated to a different patron in the Medici family, who as rulers of Tuscany also were at the head of the University of Pisa. Together these volumes treat Aristotle's *Physics, On Generation*, and *On the Soul*, all texts read as part of Pisa's course in natural philosophy. The prefatory material of the *Circulus*, its title, and its content suggest that the text was intended to celebrate Bérigard's teaching at Pisa and provide a model for his colleagues at Pisa, Padua, and elsewhere as they prepared their own lectures on natural philosophy. The surviving papers of Bellini, who taught natural philosophy and medicine at Pisa from the 1660s, testify to its use by university professors. In his reading notes Bellini cited Bérigard's *Circulus* repeatedly, in relation both to natural philosophical topics and to topics having more to do with medicine.[15]

Unlike standard teaching textbooks, which often followed the Scholastic practice of dividing Aristotle's works into *quaestiones*, Bérigard's *Circulus* is a dialogue. Bérigard employed two characters, Charilaus (literally, the "flatterer of the people"), who represented the views of Aristotle, and Aristaeus (literally "the best"), who represented the views of Aristotle's ancient opponents all rolled into one.[16] This format allowed Bérigard considerably more freedom to explore the nuances of different philosophical positions and topics in a style more appropriate for professors preparing their own commentaries and lectures than for students being exposed to the material for the first time. Through his dialogue between Aristaeus and Charilaus, Bérigard revealed the influence Pisa's eclectic approach to natural philosophy had had on his own thought. He artfully wove together both ancient and contemporary arguments, incorporating elements of Galileo's *Two New Sciences* into his teaching, where they complemented Bérigard's treatment of Aristotle's writing.

One instance occurs in the section of Bérigard's text devoted to Aristotle's *On Generation*. Here Bérigard's interlocutors begin by discussing the causes of rarefaction and accretion, as well as the relationship between these processes, heat, extension, and motion. Bérigard assigned to Aristaeus the view that rarity can be attributed to the mobility and intermixing of simple substances within a compound body. Charilaus, in contrast, relied on the behavior of the light and flame produced from an oil lamp to argue that rarity instead is a quality attributable to substances. In the course of the dialogue, Charilaus finally manages to gain some ground by forcing Aristaeus to admit that he has made a case for the occurrence of rarefaction *impropriè dicta*. In a previous attempt to reconcile the views of Aristotle and Averröes, Charilaus had made a distinction between two types of rarity. Rarity *propriè dicta* was a real quality resulting from a subtlety of the parts making up the object. In contrast, rarity *impropriè dicta* was a type of rarity attributed to materials like sponges, whose rarity was due not to the presence of any quality but rather to the separation of its solid parts by holes filled with a tenuous substance.[17]

Having forced this concession from his rival, Charilaus graciously concedes that another well-known argument for rarefaction *propriè dicta* is not very convincing. According to Charilaus, this argument is based on a mathematical demonstration that claims to show that any small surface can be made equal to any great one, a circumstance that can only occur via rarity *propriè dicta*. In particular, explains Charilaus, let a wheel with an exceptionally large diameter—he proposed the size of the last of Aristotle's heavenly spheres—be given, and let the axle of this wheel be much smaller, only one inch (fig. 5.1). As the wheel makes one revolution on plane BD, the axle will cover the same distance on the parallel plane OC. The conclusion, according to this demonstration, is that the circumferences of the very small axle and the larger wheel are equal.[18] Furthermore, claims Charilaus, according to

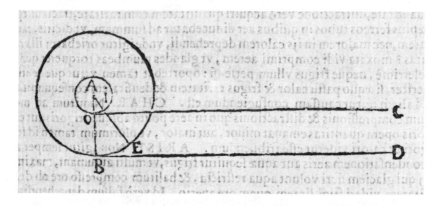

Fig. 5.1. Bérigard's rendition of the two concentric circles. "*In Aristotelis lib. De ortu & interitu,*" in Circulus Pisanus (1643), 41 (detail). Courtesy of the University of Glasgow Library, Special Collections. Shelfmark Sp Coll Veitch Eg6-d.15.

this argument, the only way for a small surface to be made equal to any greater one is through rarefaction *propriè dicta*.

Bérigard cited no source as the inspiration for this mathematical demonstration, but it is likely that it derives from Galileo's discussion of the paradox of Aristotle's wheel in Day 1.[19] Galileo included a number of diagrams to illustrate this scenario, including that shown in figure 5.2. As the similar diagrams make clear, this problem is the same as that described by Bérigard. Galileo had solved the paradox by considering the behavior of concentric polygons, which cover the same horizontal distance because the inner polygon "skips" sections of the plane as the circumference of the outer polygon continuously touches its plane. Taking the limiting case of an infinitely sided polygon (i.e., a circle), Galileo argued that the same phenomenon occurs; however, in this case, the skipped sections, which correspond to tiny void spaces, are infinitely many.[20]

After a discussion of the nature of infinities and infinitesimals, Galileo returned to his solution to Aristotle's wheel in order to offer his own explanation of the processes of rarefaction and condensation. The "mathematical demonstration" described by Charilaus parallels Galileo's proposal exactly. For Charilaus, it is the observation that the smaller surface could be made equal to the greater one that makes the assumption of rarefaction necessary. Similarly, Galileo argued that the process of rarefaction corresponded to the line described by the smaller circle while driven by the larger, whereas condensation was akin to the revolution of the larger when impelled by the smaller.[21]

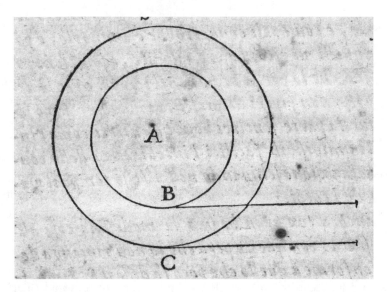

Fig. 5.2. Galileo's depiction of "Aristotle's Wheel." Two New Sciences (1638), 50 (detail). RB 701317, The Huntington Library, San Marino, California.

We can infer from this diagram and example that Bérigard read the *Two New Sciences* topically with the exegetical and referential methods of book-ish natural philosophy. Bérigard selected an excerpt from the *Two New Sciences*—that of Aristotle's wheel as it touched on the question of rarefaction—that was in dialogue with his own Aristotelian project. In his text, Bérigard folded Galileo's findings into his discussion, bringing Galileo's argument into conversation with the questions and concerns of Aristotelian natural philosophy. He did so because he had been trained in an Aristotelian tradition that embraced an eclectic approach, one that incorporated a variety of outside speculation in its exposition of Aristotle and was grounded in the juxtaposition and resolution of Aristotle's and contrary opinions.[22]

Yet by replicating Galileo's image of Aristotle's wheel, Bérigard went beyond traditional textual methods. He employed the image from Galileo's text as part of his commentary, using it as a mnemonic device associated with Galileo's claims. It served as a recognizable visual cue to readers that this argument regarding Aristotle's wheel was relevant to commentaries on rarefaction and condensation.[23] In bringing visual evidence to bear in this traditionally textual genre, Bérigard revealed himself to be methodologically versatile as well; he relied on a variety of arguments, evidence, and styles of exposition to further his presentation of natural philosophy, regardless of whether they were traditionally "Aristotelian."

Bérigard relied on this same combination of textual description and imagery to incorporate other opinions from the *Two New Sciences* into his text. In sections of his dialogue on Aristotle's *On Generation* and *Physics*, Bérigard included another image (fig. 5.3) to illustrate an experiment posed by "recent writers" for the existence of a void. According to Charilaus, these writers argued that the glass vessel BA should be filled with water. A void would then be made in the space BA when the cover C was dragged open by force.[24]

In the ensuing dialogue, Aristaeus and Charilaus discuss the merits of this proposed experiment. Aristaeus's initial reaction is to doubt that it can be carried out according to Charilaus's description. According to Aristaeus, were the proposed experiment to be tested, one would find that the water would fill the space A, because thin, aerial bodies would enter the pores of the glass, for according to Aristotle's dictum, nature abhors a void. Charilaus then proposes a modification to the original experiment whereby a heavy weight D would force the cover C to remain extended.[25] The two interlocutors agree that such an experiment would provide the opportunity to test the void, yet they also assert that the experiment would never work. Following Aristotle, Anaxagoras, and Empedocles, they conclude that a void as such would not be found in nature. Instead, thin bodies would enter through the pores of the contraption, filling the supposed empty space with matter.[26]

Like the example of Aristotle's wheel, this argument made by "recent writers" was taken from Day 1 of Galileo's *Two New Sciences*. There Galileo described a device for measuring the force of nature's repugnance to the void (see fig. 2.7). Heavy materials were placed in the bucket until the upper surface EF of the cylinder detached from the lower surface of the water. This displacement indicated that the combined weight of the iron rod, the container, and the heavy material within the bucket was the force of nature's abhorrence of the void.[27]

A comparison with Bérigard's text immediately reveals the similarities between the two procedures. Bérigard's glass vessel BA corresponds to Galileo's glass or metal cylinder CABD. Galileo's stopper, the second cylinder EGFH, is the equivalent of the cover C. Like Bérigard, Galileo filled his cylinder completely with water and hung it upside down, with the movable stopper or cover on the bottom. Both used their apparatus to effect a separation between the surface of the water and the movable stopper, or cover of the vessel, and both argued that the resulting space would be a void.

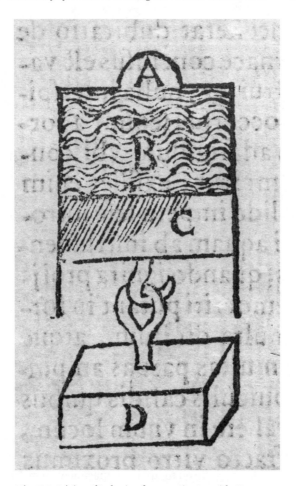

Fig. 5.3. Bérigard's device for creating a void. *"In Aristotelis lib. De ortu & interitu," in* Circulus Pisanus *(1643), 32 (detail). Courtesy of the University of Glasgow Library, Special Collections. Shelfmark Sp Coll Veitch Eg6-d.15.*

The two authors, however, held different assumptions about what their procedures indicated about the void. Bérigard portrayed his contraption as an example proposed by individuals who argued that a natural void could exist in nature. Bérigard believed that if the experimental setup were to behave as described, it would disprove Aristotle's contention that a natural void could not exist. Galileo's opinion on the void was more ambiguous. Throughout Day 1, Salviati claims that nature abhors a void and that it is precisely this abhorrence to which the resistance of materials can be attributed. The device for measuring the force of the void, in Salviati's description, is thus exactly the opposite of what Bérigard insisted: it is to measure the force by which nature prohibits a void; it is not a device for the creation of a void. Yet Sagredo offers an alternative vision, suggesting in a separate passage that other phenomena necessitate the creation of a void for an instant, so that one should not claim that such voids are unnatural.[28]

The alternative opinions expressed by Salviati and Sagredo thus offer two possible interpretations of Bérigard's reliance on Galileo's text. On the one hand, Bérigard might have read Salviati's description and applied his own interpretation to the consequences of such a device. Or perhaps Bérigard was an extremely careful reader of Galileo and found in Sagredo's comments a potentially dangerous interpretation of the apparatus that he desired to confront in his *Circulus*.

Comparisons between Bérigard's practices and those of other readers elucidate additional features of Bérigard's approach to Galileo. Like Bérigard, Mersenne read Galileo topically and found fault with his application of the paradox of Aristotle's wheel to the phenomenon of rarefaction (see chapter 3). Whereas Mersenne looked to contemporaries in order to situate and evaluate Galileo's account, Bérigard put Galileo in conversation with Aristotle. Viviani and Ward annotated the same passage as did Bérigard regarding the device for measuring the force of the void (see chapters 2 and 4). Though he had applied the bookish methods of topical note-taking and summary to other sections of Galileo's text, this passage prompted Ward to include a marginal note exhorting himself to further experimentation. In his *Trattato*, in contrast, Viviani employed this passage to comment on a later section of Galileo's text. Ward's turn to experimental replication confirms a standard narrative of Galileo's contribution to a developing experimental science and the accepted history of the reception of the *Two New Sciences*. Bérigard's response, in contrast, resembles Viviani's; both drew on established practices of sixteenth-century reading and commentary. More so than Viviani, however, Bérigard read Galileo through an Aristotelian lens, as is apparent from his decision to consider the merits of the proposed procedure within a philosophical debate concerning Aristotelian doctrine. What makes all of these readers' approaches eclectic is their willingness to explicate Galileo's text by applying to it a variety of intellectual traditions, even when those traditions seem—to modern eyes—contradictory. Labeling all these reactions "eclectic" does not make the analysis trivial. Instead it emphasizes the fact that intellectual transformation in the seventeenth century proceeded via this pluralistic approach, not through a sustained conflict between old and new.

Bérigard's conviction that novel ideas were useful to explicating Aristotle, a central feature of eclectic Aristotelianism, can be seen even more clearly in the changes he made in the second edition of the *Circulus*, published in 1661. His alterations suggest that he viewed new speculation not as an attempt to overthrow the old but as part of a quest to better answer long-established questions. One clear example of this attitude is Bérigard's decision to excise the discussion and image of Galileo's contraption to measure the force of the void (see fig. 5.3) in the sections addressing Aristotle's *Physics* and *On Generation* and replace them with a description of recent experiments with the Torricellian void. In the section of his *Circulus* that treats Aristotle's *Physics*, for example, Bérigard substituted the account of Galileo's device with the following comments by Charilaus:

> The same must be said of mercury, which is said by modern writers to leave a void space in a glass tube, and that it is made greater with fire having been brought near, and smaller with ice brought near. But since gnats fly about in it, sound is heard [in it], according to the witness Kircher, and light is produced [in it], these things demonstrate that a very subtle body is contained in that space ... evidently very thin substances are dragged through passageways of the glass.[29]

In his text, Bérigard referenced the recent experiments with inverted tubes of mercury, first carried out by Galileo's student Evangelista Torricelli. Torricelli had found that when a glass tube was filled with mercury and then inverted in an open dish, the mercury in the tube always fell to the same level. Reports of Torricelli's experiments inspired other natural philosophers across Europe to carry out their own trials. They were amazed to find that the level of the mercury changed with altitude but that it remained unaffected by variations in the width and inclination of the tube. Most puzzling was the space above the mercury, which many, following Torricelli, interpreted to be a void space or vacuum.[30]

In his *Circulus*, Bérigard gave little information about how such experiments were carried out, but he did provide details, including the facts that gnats can fly in the space, that light can be produced in it, and—citing the experiments of the Jesuit mathematician Gasparo Berti (ca. 1600–1643) reported in Athanasius Kircher's (1601/2–1680) 1650 *Musurgia Universalis*—that sound is heard in it.[31] These phenomena, argues Charilaus, are evidence that the space above the mercury is not empty but in fact contains very thin substances that enter through passages of the glass.[32] With this comment, Bérigard explained the new mercury experiments using the same reasoning he had employed to dismiss Galileo's device.

Bérigard's decision to replace Galileo's apparatus with a reference to the mercury experiments further disrupts the narrative of conflict by revealing his own familiarity with the state of natural philosophical investigation. Substitution of Galileo's proposed procedure with a description of the recent mercury experiments was a sound decision, because both experiments were understood as being ways to create a void, thus violating Aristotle's provision that a natural void was impossible. They were also related because Galileo's apparatus was the intellectual inspiration for the latter. Torricelli had first filled tubes with water following Galileo's description in his *Two New Sciences*. He had then turned to filling the tubes with the heavier liquid mercury, which had the advantage for experimenters of having a much lower maximum height than did water.[33] By replacing his description of Galileo's apparatus with reference to these mercury experiments, Bérigard thus updated his text by reporting on the next generation of experimental attempts to succeed Galileo's proposal.

Bérigard's attitude toward established and novel hypotheses in the second edition is akin to his attitude in the first. Though he rejected the vacuist interpretation of the mercury tubes, he accepted both the premise of experimentation and the reported phenomena. He believed, moreover, that both were relevant to his explanation of Aristotle. What is most interesting is the fact that Bérigard edited the passage on Galileo to keep his text abreast with the most up-to-date speculation. This change indicates Bérigard's view that the old and the new are in a continually evolving dialogue, one in which the latest experiments should be brought to bear on the discussion. In contrast to the rhetoric of his successors at Pisa, Bérigard's practices reveal him to have seen old and new as compatible and complementary. According to Bérigard, one could embrace parts of the new without abandoning the Peripatetic enterprise. Such an attitude was longstanding in the Aristotelian tradition, which since antiquity had seen the incorporation of outside philosophical doctrine. However, the point is worth emphasizing given contemporary rhetoric at Pisa underscoring the incompatibility of old and new approaches.

Bérigard's Topical Reading in Context

The topical and selective reading observed in Bérigard's teaching of the *Two New Sciences* follows a pattern observed both in Bérigard's reading more generally and in that of other university professors. It is confirmed, first, by Bérigard's treatment of Galileo's *Dialogue*, a text he knew well. To compose his *Dubitationes*, Bérigard read Galileo's 1632 publication with some care, and he referenced it in other sections of his *Circulus Pisanus*. In his discussion of Aristotle's *On the Heavens*, for example, Bérigard's interlocutors consider the question of the Earth's immobility at length, and they specifically cite Galileo's claims in the *Dialogue* that the motion of the sunspots and the tides can be explained as a result of the Earth's motion.[34] Despite this familiarity, Bérigard nowhere in his *Circulus* considers the *Two New Sciences* and the *Dialogue* together. Bérigard's treatment of Galileo's final publications thus proceeded in two independent directions. He read the *Two New Sciences* in light of Aristotle's writings on rarefaction and the void, but he turned to the *Dialogue* to illuminate his discussion of Aristotle's *On the Heavens*. This pattern indicates that Bérigard read not for Galileo's overall argument but selectively and topically, seeking out excerpts that would help him address questions central to his exposition of Aristotle.

Bérigard responded similarly to the writings of another contemporary, William Harvey (1578–1657). Whereas Galileo remains unnamed in his text, Bérigard mentions Harvey by name in the sections of his *Circulus Pisanus* that treat Aristotle's *On the Soul*. Though he did not endorse Harvey's theory of circulation, Bérigard discussed Harvey's examination of insects with microscopes and admitted Harvey's experimental finding that blood springs from punctured arteries and the heart. Bérigard, moreover, adopted both Harvey's argument that the active phase of the heart's motion is systole and his claim that the arterial pulse is synchronous. As in the sections in which he discussed Galileo, Bérigard's purpose was not to endorse or refute Harvey's theories but to expound on Aristotle by drawing on both old and new textual sources.[35]

Selective and topical reading strategies were a common practice employed by other university professors elsewhere in Europe, many of whom incorporated the speculations of innovators in their teachings of Aristotle. Historians have described a process of appropriation whereby Aristotle's writings provided the lens through which early modern professors sifted contemporary publications. According to Laurence Brockliss, by the third quarter of the seventeenth century professors had largely reformulated the content of the traditional physics course in light of the new work done in astronomy, physiology, dynamics, and pneumatics. Citing the highly popular textbook of the Paris professor Pierre Barbay (d. ca. 1675) first published in 1675, Brockliss described the content of Barbay's course as "unquestionably" Aristotelian, even though it embraced certain novel ideas, including Tycho Brahe's (1546–1601) geoheliocentric system and Galileo's telescopic discoveries.[36]

Galileo's work on mechanics and motion was among the novelties certain professors across Europe incorporated into their classroom teaching. In the Netherlands, for example, the introduction of Galileo's ideas on motion into the classroom happened relatively early. In his 1639 *Disputatio*, Jacob Ravensberg, later professor of mathematics at Utrecht, affirmed Galileo's conclusion that all heavy bodies fall with the same speed in his discussion of the Copernican system. Just over a decade later, Johannes Phocyclides Holwarda (1618–1651), then professor of philosophy at Fraeneker, discussed Galileo's proposition that projectiles follow a parabolic trajectory in his *Philosophia Naturalis*, a publication that presented the contents of his classroom teaching.[37] Galileo's rules for the vertical fall of heavy bodies, their motion along inclined planes, and the behavior of pendulums were included in the published lecture notes of Rinaldini, who taught at Parma, Pisa, and Padua in the second half of the seventeenth century.[38] Reference works read by professors of mathematics in Spain, including Juan Caramuel y Lobkowitz's (1606–1682) 1670 *Mathesis Biceps Vetus et Nova*, described Galileo's results on the isochronism of the pendulum and his rule of fall.[39] Students in France learned of the findings in Galileo's *Two New Sciences* through such textbooks as Emmanuel Maignan's (1601–1676) 1653 *Cursus Philosophicus*.[40]

When these professors appropriated elements of the New Science into their teaching, they folded into portions of the curriculum passages in which Galileo and other authors could be seen to speak to established questions. Like Bérigard, they read topically and selectively. One example is Rinaldini, who organized his *Naturalis Philosophia* around established questions and topics of Aristotelian natural philosophy. New material was incorporated within this framework. Rinaldini thus included discussions of the inverted mercury tubes in a section of the text dedicated to the question of the void.[41] Galileo's findings on local motion were inserted within traditional questions about the nature and cause of motion and within discussions of how to measure space and time.[42] Ugo Baldini has claimed that Rinaldini's *Naturalis Philosophia* allows a glimpse into how "a contemporary of elevated philosophical and scientific culture perceived the Scientific Revolution." In Baldini's analysis, Rinaldini was a unique figure, one whose deep understanding of both the Aristotelian tradition and that of the New Science allowed him to see and explore their continuities.[43] Placing Rinaldini in the context of Bérigard's teaching, however, suggests that while Rinaldini may have been exceptionally well versed in the two traditions, his approach was decidedly commonplace. Professors like Bérigard and Rinaldini, trained to study Aristotle eclectically, saw Galileo's work as speaking to the traditional questions and topics of Aristotelian natural philosophy. They saw nothing amiss in bringing Galileo into their commentaries on Aristotle nor in applying the techniques of traditional reading to Galileo's book.

Mining the *Two New Sciences* for Practical Mathematics

Bérigard's *Circulus* suggests that he approached the *Two New Sciences* with the methods and goals of traditional natural philosophy, reading selectively and topically in order to explicate Aristotle. The surviving manuscript teaching notes of his near contemporary Vincenzio Renieri reveal how other discipline-specific scholarly practices translated into a different use for the text in the classroom. Renieri came to know Galileo only in 1633, after his condemnation. When Dino Peri's (1604–1640) death left the mathematics chair at Pisa open, Galileo lent his support to Renieri, who was named to the post.[44] Renieri's surviving manuscript writings include a number of treatises that appear to have been intended for teaching. These manuscripts include one titled "On the Analysis of Triangles," fragments and drafts from treatises on practical geometry, various prefaces to Euclid, and two chapters from a treatise titled "On Fortification." Renieri also composed notes on astronomy, including multiple versions of a manuscript entitled "De Sphaera Mundi," as well as writings on the construction and use of gnomons.[45]

Renieri addressed Galileo's *Two New Sciences* in only one of these manuscripts, his undated treatise on practical geometry, most likely composed between 1638 and 1647. Unlike his "De Sphaera," which would have been used in formal settings, such as a university lecture or disputation, Renieri's manuscript treatise was probably employed for the instruction of private pupils. He composed it in the vernacular, not Latin, the language of the public university, and he dedicated it to practical geometry, a subject not included in the public cycle of lectures.

Renieri's treatise describes the construction of the geometrical quadrant, an instrument used by surveyors and navigators to measure angles of up to 90 degrees. It then provides a series of problems detailing how the quadrant could be used to determine unknown lengths and distances. Renieri divided the text into four chapters. The first three are devoted to various methods of determining unknown lengths oriented horizontally, perpendicularly, and obliquely with respect to the horizon; in the fourth, he considers the measurement of surface areas. Though the treatise is dedicated to the use of the quadrant, Renieri also has his reader use mirrors and other devices in place of the quadrant in some problems. This type of exercise was common in the early modern period. Mathematicians, including Galileo himself, estimated the heights of towers using gnomons, mirrors, and eventually the compass.[46]

Renieri cited results from the *Two New Sciences* in a problem that asked the reader to determine an unknown vertical height using a pendulum (fig. 5.4). Its title reads, "To measure a height, though it cannot be seen, by means of a cord falling from the top of the height, only whose lower end is seen."[47] Renieri imagined a tall construction, in this case a tower, from which was hung a long string or cord AB. While the person making the measurement could see the lower end of the cord (at B), he could not see the top of the tower or the top of the cord.

Renieri began his solution to the problem by referring to Galileo's observations of the behavior of pendulums, noting that "this very noble Problem depends on the noble observation of Mr. Galileo Galilei regarding the oscillations that heavy objects attached to some cord make when they are moved from the perpendicular."[48] Renieri summarized Galileo's findings with the help of another diagram (fig. 5.5). He explained that according to Galileo, the ratio of the length of the cord EF to that of AB was equal to the squares of the ratios of the number of vibrations made in equal times by B and F. Renieri then described how the construction of a smaller pendulum of known length, which he designated CD, could be used to find the length of the long cord and thus the length of the tower. His procedure involved treating the long cord hanging from the tower (AB) as a pendulum, which was made to swing between points F and E. The length of the cord AB and thus of the tower was found by using the relationship Galileo proposed between the period and the length of the pendulum. Renieri told his reader to count the number of oscillations completed in the same amount of time by the long cord AB and the shorter pendulum CD. This ratio, when squared, would be equal to the

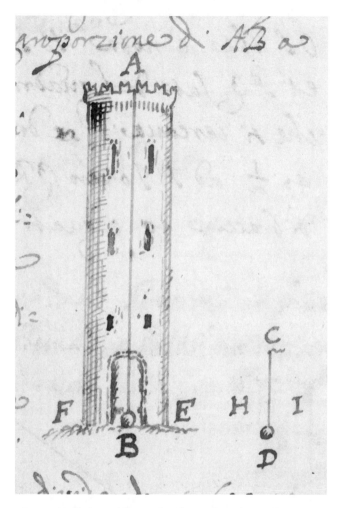

Fig. 5.4. Renieri's problem to determine the unknown height of a tower. *BNCF Gal. 114, fol. 47r. Courtesy of the Ministero dei beni e delle attività culturali e del turismo / Biblioteca Nazionale Centrale di Firenze.*

inverse proportion of the lengths of the two cords. The unknown length AB of the cord and tower would be calculated from the known length of the shorter pendulum (CD).[49]

Galileo himself had employed the conversation between his interlocutors to suggest such a use for his text. In response to Salviati's description of the relationship between the length of a pendulum and its vibrations in Day 1, Sagredo comments that such results could be applied to the well-known

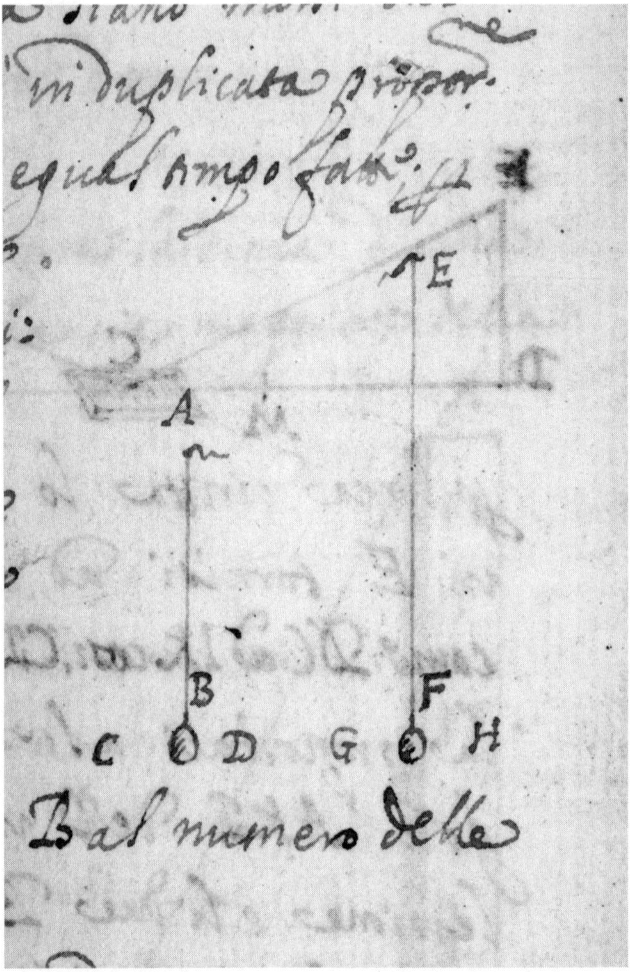

Fig. 5.5. Renieri illustrates Galileo's findings on the pendulum.
*BNCF Gal. 114, fol. 46v. Courtesy of the Ministero dei beni e delle
attività culturali e del turismo / Biblioteca Nazionale Centrale di
Firenze.*

genre of problems involving finding the unknown height of a tower. According to Sagredo, "I can easily know the length of a string hanging from any great height, even though the upper end of the attachment is out of my sight and I see only the lower end." Sagredo then goes on to explain the same procedure outlined by Renieri of employing a second pendulum. He follows this description with an illustrative calculation.[50] The close correspondence between the two suggests that Renieri devised his problem directly in response to Galileo's text.

Renieri's turn to the *Two New Sciences* to seek out sample mathematical problems resembles the approach of pseudo-Baliani, Viviani, Ward, and Wren, who annotated many parts of the book as if it were a collection of mathematical exercises. Like these annotators, Renieri saw Galileo's text as a means of furthering mathematical prowess. The problem he posed to his students was less a demonstration of Galileo's physical findings than a display of the various possible permutations relating to the practical geometrical exercise of measuring unknown distances. Renieri's text suggests what the marginalia of pseudo-Baliani, Viviani, Ward, and Wren might have become if they had left the margins of the printed book and entered a pedagogical treatise.

Like Ward, Wren, Viviani, and Bérigard, Renieri also read, or at least taught, the *Two New Sciences* separately from the earlier *Dialogue*. As a professor of mathematics, he was responsible for delivering lectures on astronomy, and in this capacity he presented material from Galileo's *Dialogue*. His chapter titled "On the Earth's Rest" in his treatise "De Sphaera" considered various arguments for the Earth's motion put forward by Copernicans. Here Renieri described Galileo's theory of the tides, found in Day 4 of the *Dialogue*, without attributing it to Galileo.[51] He also countered standard Aristotelian-Ptolemaic arguments that the behavior of projectiles on the Earth's surface offered evidence of the Earth's rest by describing the behavior of fish in a bowl and flies in a closed room of a moving ship, an example also described in Day 2 of the *Dialogue*.[52] At no point in his treatise, however, did Renieri bring up arguments from the *Two New Sciences* or allude specifically to Galileo's quantitative findings on motion. As Renieri taught Galileo's *Dialogue* and *Two New Sciences* to his students, the former was a work pertaining to astronomy, while the latter was a work relevant to mixed mathematics.

Developing a Quantitative, Causal Physics

The surviving notes of Pascasio Giannetti offer an example that contrasts with those of Bérigard and Renieri. Giannetti attended Pisa as a student before being hired to teach logic, natural philosophy, and medicine from 1682 to 1737. According to the eighteenth-century historian Antonio Fabroni, when Giannetti refused to rein in his discussion of novel doctrine in his courses on natural philosophy, he was made the chair of theoretical medicine and forced to teach natural philosophy only privately.[53] Giannetti, unlike his predecessors at Pisa, read and taught the *Two New Sciences* in tandem with the *Dialogue*. More so than Bérigard and Renieri, he also drew on a set of diverse reading practices influenced by new currents in philosophical, experimental, and mathematical investigations. Other aspects of his approach, however, were derived from more traditional natural philosophical practices and resemble those of Bérigard.

Certain features of Giannetti's extant manuscript notes indicate that he, like Renieri, employed them precisely in a private setting. The title page of his manuscript contains the attribution "The *Physics* of the very distinguished man Pascasio Giannetti, Professor of Public Medicine and of Philosophy at the Pisan Lyceum."[54] This designation of Giannetti as professor of public medicine—he was not appointed until 1706—suggests that these notes were used in private lessons only after Giannetti became the chair of medicine. Like other university professors, Giannetti most likely relied on his text for several years as he retaught his course. He divided the pages of his notes in two and wrote only on the right side. Occasionally he inserted additional comments and citations on the left. In addition, he apparently returned to amend his notes in several instances. In explaining Cartesian vortices, for example, he initially spelled *vorticem* (vortex) with an *e* in place of the *o* (*verticem*) but then corrected it on the left.[55] These signs of reuse could suggest as well that these notes were originally used for his public teaching of natural philosophy prior to his appointment as professor of medicine.

Giannetti's teaching indicates that he adopted an eclectic approach that moved him beyond the immediate practices and goals of Aristotelian natural philosophy. According to his preface, the differing and contradictory opinions of philosophers had made natural philosophy "very difficult and so confused" that it was an "immense job . . . to record those things that have been agreed upon and to judge them." As a result, Giannetti told his readers, he did not follow Aristotle directly but chose "the most useful" of philosophers' opinions and selected explanations that were "both short and easy for philosophizing."[56] Some of the topics Giannetti chose to treat in his course figured little, if at all, in traditional teaching, including "On the Motion of Projectiles" and "On Reflected Rays."[57]

Giannetti engaged significantly with Galileo's findings on motion in two sections of his text. Parts of his discussion appear to derive from a reading of Day 2 of the *Dialogue*, while others likely drew from the *Two New Sciences*. However, because Giannetti followed his own advice to provide explanations and principles that were "short and easy" for philosophizing, for the most part he did not indicate his sources with precision, eschewing the type of textual attribution associated with Scholastic and humanist scholarship. Giannetti often cited few or no textual authorities, and he did not discuss at length the contrary opinions of various authors. This pattern suggests that he employed and modeled for students different scholarly practices than did Bérigard.

In the section of his text entitled "On Accelerated Motion," for example, Giannetti offered an exposition of Galileo's science of motion, complete with diagrams, with little discussion of sources or controversies. He noted at the outset that while it had always been observed and accepted that heavy bodies falling downwards fall with greater velocity, the exact ratios by which velocities and spaces were augmented were not known before Galileo. Giannetti then set forth Galileo's odd-number rule, which he explained with the use of a diagram (fig. 5.6). The odd-number rule states that if a heavy body F departs from rest at point A and descends through the line AG, it will run through the space AB in the first interval of time. In the second, equal interval of time, it will run through BC, which will measure three times that of AB, and so forth. Students were told to trust the rule because it had "been tested by innumerable experiments" and was "accepted by all," but no other text or justification was offered.[58] Though this result was presented formally in Day 3 of the *Two New Sciences* and described briefly in Day 2 of the *Dialogue*, Giannetti gave his students no indication of his source for the rule. By teaching this way, Giannetti modeled a new type of reading practice. Rather than amassing textual authorities and opinions, his students were to focus on quantitative relationships and diagrams.

Giannetti also employed a diagram to illustrate a second quantitative rule presented in both the *Dialogue* and the *Two New Sciences*. This second rule stated that the distances traveled by a heavy body are proportional to the squares of the time. Again Giannetti offered no accompanying mathematical proof, specific experimental evidence, or textual attribution. Instead, the reader was to accept Galileo's statement on the basis of the previously cited odd-number rule, the constructed diagram, and the definition of the square of a number. According to Giannetti, this second rule "is understood from the explanation alone of the expressions or names," that is, of the definition of the square of a number. Assuming that the distance AB measures a cubit, Giannetti concluded that the movable traveled a distance of one cubit

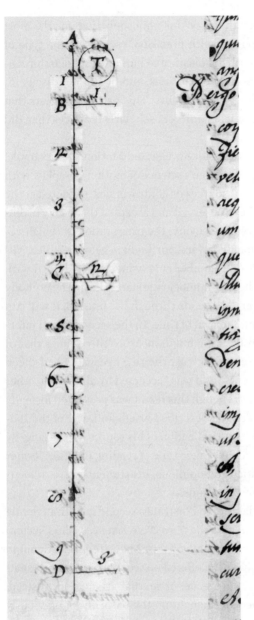

Fig. 5.6. Giannetti's diagram to explain Galileo's odd-number rule. *Firenze, Biblioteca Medicea Laurenziana, MS Ashb. 872, c. 32v. Courtesy of the Ministero dei beni e delle attività culturali e del turismo.*

at the end of the first unit of time (one is the square of one), a distance of four by the end of the second unit (four is the square of two), a distance of nine in the third (nine is the square of three), and so on.[59]

Giannetti followed this explanation of Galileo's results by presenting applications likely drawn from the *Dialogue*. At one point, for example, Giannetti corrected a specific passage in the *Dialogue* claiming that a one-hundred-pound ball will fall one hundred braccia in five seconds. Giannetti labeled this calculation an "error." He argued that "accepted opinion proved by very certain experiments" showed that the space traversed by an iron ball certainly exceeded twelve hundred braccia.[60] Giannetti offered no source for his "accepted opinion" or "very certain experiments," but both Baliani and Mersenne, among others, had found this passage problematic and attempted to correct it by experiments.[61] This correction was followed by a brief allusion to the way the odd-number rule could be applied to calculate the falls of bodies along inclined planes and the motion of pendulums.[62]

Galileo's findings on motion, drawn from both the *Dialogue* and the *Two New Sciences*, were also the focus of the next section of Giannetti's teaching text, entitled "On the Motion of Projectiles." Giannetti here began by asserting that gravity or heaviness always acts in violent motion according to the rules he outlined earlier. To illustrate this concept, Giannetti told readers to imagine sitting or standing on a moving ship. If the ship stopped quickly, they would feel an impressed motion, which would make them fall. Citing an example treated in Day 2 of the *Dialogue*, he explained that heavy bodies dropped from the mast of a quickly moving ship fall to the foot of the mast in time with the ship's motion.[63]

Unlike in the earlier sections, Giannetti here referred readers to recently published texts. He described experiments carried out by Pierre Gassendi and detailed in his 1640 *De Motu*, which involved dropping stones from the mast of a quickly moving vessel.[64] His citation of Gassendi likely explains why Giannetti relied on the *Two New Sciences* and the *Dialogue* in tandem. While Gassendi's discussion itself focused largely on results and scenarios presented in the *Dialogue*, he explicitly explored the links between Galileo's science of motion and his cosmological commitments. As a result, Giannetti may have been less tied to discipline-specific modes of reading than his predecessors, who had tended to associate the *Dialogue* with discussions pertaining to astronomy and the *Two New Sciences* with other areas of natural philosophy. Indeed, following his discussion of the ship, Giannetti cited Galileo's *Two New Sciences* indirectly when he outlined more quantitative information regarding the path traced by projectiles and referred readers to unspecified works by Galileo and Torricelli.[65] Here Giannetti focused on Galileo's results but omitted many of the mathematical definitions and propositions Galileo employed to prove them. According to Giannetti, Galileo demonstrated that the horizontal aspect of projectile motion is equable or has constant velocity. The vertical aspect of the motion, however, is determined by heaviness (gravitas), which always acts in accordance with the odd-number rule.

Giannetti suggested that readers interested in the curve described by a projectile pick up a ruler and compass—traditionally the tools of a mathematician, not a natural philosopher—and draw out the line themselves. He instructed readers to construct two straight lines at right angles to each other. One line should be divided into parts according to the odd-number rule. The second should be divided into as many equal parts as the first. Parallel lines should then be led from these divisions, and the points at which they intersect would indicate the path taken by the projectile. Giannetti declared that these intersections would make a curve called a parabola.[66] Unlike in other sections of his text, Giannetti did not include a diagram to illustrate this construction. It is possible that he led students through the construction himself or that he merely mentioned the possibility and left the drawing of the curve to the students.[67]

Giannetti's presentation offers clues regarding his own reading methods and the scholarly practices he hoped to impart to students. His students of natural philosophy were to be proficient to some degree in the tools of the mixed mathematician so that they might construct the parabolic trajectory of projectiles. At the same time, they did not need to be familiar with mathematical proofs, concepts, and definitions of the type Galileo employed in the *Two New Sciences*. Nor did students need to know how to use mathematical tables or how to carry out more complex calculations, for Giannetti provided no information on how to compute altitudes or ranges of projectiles. Instead, he noted only that the knowledge of the parabolic trajectory was of "great use" to soldiers. As a result, he reported, tables had been constructed by other authors relating the angle of projection to the distance traveled by the projectile.[68] While Giannetti referred interested students to the writings of Galileo and others, his teaching notes suggest that for him, all that mattered was that his students be capable of citing and describing Galileo's basic findings. In this part of his teaching treatise, textual comparisons and citations, the central activity of bookish natural philosophy, were largely absent. Giannetti read for correct answers and their applications, not for the accumulation of multiple opinions for discussion and debate. In these two sections of his teaching treatise, Giannetti read and taught the *Two New Sciences* using methods that resemble those of a practical mathematician, but for an enterprise that he identified as natural philosophy or physics. His impulse to pronounce on incorrect passages in the *Dialogue* and to read the *Dialogue* and the *Two New Sciences* in tandem also reveals his adherence to some of the traits of the "ideal" reader explored in chapter 1.

Yet, Giannetti did not completely abandon the methods and concerns of traditional philosophy. In this sense, he marked himself as a reader similar to the annotators discussed in chapters 1–4, who read the *Two New Sciences* using an eclectic set of methods and with diverse goals in mind. In another section of his treatise, entitled "On Heaviness," he combined traditional and novel approaches more thoroughly. Here he began not with quantitative results but with Aristotle. Giannetti noted that the material of the Earth is divided into different types or elements—earth, water, air, and fire—that all move in different ways and mix together. In agreement with Aristotle, Giannetti claimed that it was appropriate to seek the origin and cause of these motions, since they were at the heart of many natural phenomena. Furthermore, because these motions were directed and not all elements existed in a chaotic mass, it could be concluded that there was some strength or principal mover directing the elements. According to Giannetti, Aristotle agreed with this assessment and, recognizing the twofold nature of motion, claimed that there were two such moving principles. The first, heaviness, directed bodies to the center of the Earth; the second, levity, led bodies away from it.[69]

Giannetti found this account wanting, and he corrected Aristotle, first by employing logical arguments and textual citations. His primary objection was that nature would have wanted the Earth to be conserved. Heaviness, he argued, helps in this regard, but lightness does not, since levity incites terrestrial elements to separate from the Earth. Attempts to argue for nature's demonstration of lightness, furthermore, amounted to little, for what had been observed of the ascent of supposedly light bodies always occurred when these bodies were surrounded by heavier bodies.[70] In customary humanist and Scholastic fashion, Giannetti drew on textual sources to reinforce his logical reasoning. In the margins of his notes, he cited both ancient and contemporary sources, noting that "even Democritus and Plato confirm our opinion" and that "if there is anyone who does not profess that he is defeated by our reasoning, Giovanni Alphonso Borelli is here with his book *De mo-tionibus naturalibus a gravitate pendentibus*."[71]

Giannetti went on to provide experiential evidence to strengthen his argument, citing both contemporary experiments and those reported by ancient authors. All metals, gold being the exception, stated Giannetti, will rise when immersed in mercury, yet no one argued that these metals have levity. Instead their ascension is the result of the thrusting out of the heavier mercury. Referring to Robert Boyle's experiments with the air pump, Giannetti cited the additional example of smoke, which ascends in air but descends in evacuated glass vessels. Experiments to measure the heaviness of air also received mention. According to Giannetti, Aristotle had argued that the air around us has weight, but his followers had denied his claim.[72] Yet more recent thinkers—he did not mention names—had shown that glass vessels are heavier when filled with air than when not.[73]

Giannetti returned to the concerns of traditional natural philosophy when he explored a central concern of the Aristotelian enterprise, namely, the search for the nature and cause of heaviness. In addressing this question, however, he turned to modern sources, summarizing and criticizing the opinions of Descartes and Gassendi. Descartes's theory posited the Earth's motion, an assumption Giannetti declared was contrary to faith and reason.[74] Gassendi's explanation, according to Giannetti, had been shown false by arguments made in Borelli's *De Vi Percussionis Liber*.[75]

In this section of his text, Giannetti brought together a mix of old and new sources, methods, and goals to answer a question he saw as common to all, namely, what is the origin and cause of the diverse motion of terrestrial elements? He worked with both ancient and contemporary sources to answer this question, and he relied on a set of methods, including textual citation, logical argument, and experimental evidence, that he shared with his sources. Old and new authors were placed in conversation in Giannetti's text because they all addressed the same question and because they responded to it using similar types of evidence. Giannetti's approach can be seen as a more drawn-out and formal version of pseudo-Baliani's annotation of Galileo's discussion of the cabbage leaves. Pseudo-Baliani had corrected Galileo's account of the phenomenon by turning to Aristotelian doctrine, mechanical philosophy, and the traditional genre of pedagogical dialogue. We see in Giannetti a similar move, as he gathered together the writings of Aristotle, the opinions of ancient, medieval, and contemporary writers, and experimental evidence to address questions of heaviness and motion.

Giannetti's teaching text reveals that while he appreciated Galileo's quantitative approach and saw it as an integral part of his natural philosophy, he simultaneously assigned a place to Aristotelian causal methods and explanations, modified as appropriate in light of recent speculation and experimentation. Giannetti's arsenal of scholarly practices, which were employed in the realm of what he understood as natural philosophy, included both traditional bookish approaches and new methods advocated by Galileo and similar to those used in the realm of practical mathematics. Giannetti's text makes clear that in practice, period actors saw the choice between old and new methods as neither clear-cut nor immediate. Philosophers did not wake up one day and decide to don "Galilean" or "Aristotelian" hats; rather, they picked and chose from the different approaches espoused by ancient, medieval, and contemporary sources, and they often saw complementary and mutually supporting elements in approaches, conclusions, and texts that to modern eyes look contradictory.

Lorenzo Bellini's Commonplace Notes and Pisan Reading Practices

Professors well versed in the arguments of new philosophers put old and new methods and texts in dialogue in their writings as a result of the reading and writing practices they employed. This was the case even when they, like Giannetti, largely abandoned Aristotle in favor of new approaches. The surviving notes of one Pisan professor suggest that their practices continued to be shaped by the traditional note-taking and reading strategies undergirding humanist and Scholastic scholarship. These strategies favored close reading, an attention to textual authorities, and identifying commonalities in disparate works.

Bellini taught natural philosophy and anatomy at the University of Pisa in the second half of the seventeenth century. Having studied under self-proclaimed followers of Galileo, including Borelli and Marchetti, he himself achieved wide acclaim during his lifetime for his publications on anatomy, which applied the latest experimental and physicomathematical techniques to explain the structure and function of the kidneys and the sense organs. Bellini's own writings indicate his familiarity with a wide variety of ancient and early modern authors, from Democritus and Anaxagoras to Galileo, Gassendi, and Descartes.[76]

While Bellini read, discussed, and oftentimes embraced novel hypotheses in his publications, he continued to employ traditional textual practices in his working methods. His extant papers contain three different collections of reading notes taken in a style that resembles the commonplace method, the approach to note-taking advocated by humanist pedagogues.[77] Bellini's commonplace notes follow a slightly different format than that described by Erasmus and other sixteenth-century humanists. Rather than devoting each page to a different topical heading, as earlier humanists had advised, Bellini organized his notes loosely topically and alphabetically. The initial entry on each page corresponds to a specific subject, such as "Mixtio" or "Motus" or "Anima," but the subsequent entries often address a variety of topics, all of which begin with the same first letter. Thus on the page with the initial entry "Anima" (fig. 5.7), Bellini took notes on a number of topical heads beginning with the letter *a*, from "Aqua," to "Aer" to "Anima mundi." Bellini recorded entries on these topical headings nonconsecutively. In the section under the heading "Anima," Bellini included entries on air (*aer*) and water (*aqua*). He noted, for example, that Aristotle had asserted in the fourth book of his *On the Heavens* that water was heavy. Bellini then turned to the subject of air, indicating both Aristotle's and Plutarch's opinions on its qualities. Reverting back to water, he noted various opinions of Empedocles, Strato, and Gassendi.[78]

This organization bears some resemblance to the "New Method" of commonplacing described by John Locke in his *Bibliothèque universelle et historique* of 1686, in which an alphabetized index served as a finding aid to Locke's collection of commonplace notes. The first entry made according to Locke's method was written on the first available page. The commonplace head was entered in the left margin, with the entry accompanying it placed alongside the subject heading, leaving the left margin free. The number of the page on which the entry was made was then written in the alphabetical index at the front of the book. Subsequent entries in the same alphabetical category as entries recorded previously were recorded on the same page as the previous entries.[79] Like Locke's, Bellini's notes are taken in the commonplace style because each individual note is classified by its topic. The commonplaces themselves are organized alphabetically according to the first letter of the topical heading, but Bellini included no finding aid, apart from alphabetized tabs, in his notes. Because Bellini's extant papers do not contain any methodological statements or other explanations of his note-taking processes, it is unclear whether these notes represent his original reading notes or some intermediate or final stage.

Bellini's juxtaposition of ancient writers and Gassendi in his notes on water highlights a more general feature of his papers, namely, his tendency

183

Fig. 5.7. A page of Bellini's reading notes. *Firenze, Biblioteca Medicea Laurenziana, MS Ashb. 638.4, c. 183r. Courtesy of the Ministero dei beni e delle attività culturali e del turismo.*

to put old and new sources in dialogue. Bellini's reading notes reveal his perception that these ancient and modern authors all spoke to standard, shared topics of interest to his own scholarship. The entries under "Motion and Mover[s]" demonstrate more fully this aspect of Bellini's reading.[80] In this section of his notes, Bellini collected the ideas of a diverse group of authors, from Aristotle to Galileo. As was the practice in commonplace books, Bellini jotted down the ideas of different authors, even those whose approaches or conclusions seem contradictory to modern sensibilities. Bellini, for example, described and provided textual references for Aristotle's distinction between natural and violent motion, as well as Aristotle's notion that all motion was the result of the action of a mover.[81] Bellini interspersed these entries with Galileo's quantitative rules on the behavior of accelerated and projectile motion.[82] Similarly, he juxtaposed Galileo's quantitative rules for accelerated motion to Gassendi's speculations on the cause of such motion, even though Galileo in his *Two New Sciences* had dismissed such queries as extraneous to his quantitative project.[83]

Bellini's commonplace notes reveal that despite his embrace of novel ideas in his published writings, he relied on and modified traditional reading and note-taking methods. For all that Galileo may have claimed at the beginning of Day 3 of the *Two New Sciences* that he was putting forward a whole new science concerning a very old subject—a reference to the primacy of motion in Aristotle's writings and to the very different quantitative and experimental approaches he himself espoused—readers like Bellini applied to Galileo's work the very methods that Galileo, in his rhetorical proclamations, professed to have rejected.

Rhetoric versus Scholarly Practice

Reconstructing the goals and methods of professors' reading based on printed and manuscript teaching texts reveals a concerted attempt to engage with Galileo's ideas by applying both established and new reading practices to them. Such efforts contrast with the dominant image of Pisa as an institution either in decline or plagued by competing factions. Marchetti, Bellini, and Borelli described a group of Peripatetics so committed to Aristotle that they openly disregarded sensory experience, the writings of others, and even the basic rules of courtesy in Aristotle's defense. While extant teaching notes reveal little about how professors treated one another, they do suggest that Marchetti's bold statement that his Peripatetic colleagues valued Aristotle's *ipse dixit* above all else did not hold. While professors like Bérigard did prefer to maintain Aristotle's conclusions if possible, they did not "laugh, mock, and tease all the others." Instead, they were willing to engage with new speculation, accept and consider experiential and even experimental evidence, and at times revise long-held views.

The discrepancy between the practice and the rhetoric of Pisan professors suggests that both groups tended to exaggerate the extent of their intellectual divisions. Such a conclusion is especially persuasive given other indications that the driving concerns of many Pisan professors involved in the controversy centered on appropriate pedagogy and authority, not philosophical approach. The comments regarding differences in philosophical methods make up only a small portion of the extant letters generated during the conflict. These letters reveal that in defending the teaching of Aristotle at Pisa, Maffei was preoccupied by issues of authority, historical precedent, proper pedagogical practice, and the maintenance of religious and political tranquility. Maffei, in fact, began his letter to the grand duke by noting that his colleagues, in teaching the latest authors in place of Aristotle, defied both university statutes and precedent at Pisa and elsewhere, for the statutes in the past had been "punctually observed by outstanding philosophers, that is, Chiaramonti, Sevieri, Talentoni, Bellavita, Cintoletta, Bérigard, Marsili, and Rinaldini." Furthermore, he noted, universities all over Italy continued to teach publicly the Peripatetic doctrine.[84]

Maffei returned repeatedly to the difference between public and private teaching. To strengthen his argument that public teaching required adherence to traditional sources, he quoted from a 1646 work on Democritus written by the French atomist Jean Chrysostôme Magnen (ca. 1590–ca. 1679). According to Maffei, Magnen had told his readers that despite his own atomist tendencies, he felt that "for the education of youth it is better to read Aristotle first." Though Magnen had written a book about Democritus, he assured his readers that he always based his public lectures on Aristotle. New doctrine was to be reserved for private study; public lectures were for education, not for indulging in novel predilections.[85] This opinion—that a sound grounding in old doctrine was necessary for undergraduate students—was one commonly held in the period, even at purportedly more open institutions like the University of Oxford and by individuals, including Locke, who embraced new learning.[86]

Teaching novel doctrine in the public lecture hall did more than infringe on abstract rules drawn up in dusty university statutes. It also decreased Maffei's and his colleagues' status and prestige in ways that were tangible.

When professors publicly challenged Aristotle in the lecture hall, in their writings, and even in print, they neglected to consider the *gravità magistrale*, or the weight of their position as professors. From this neglect, Maffei claimed, was born students' contempt of their own teachers. Disputations in which participants disagreed about the fundamentals gave rise to a great desire to continue debating, but these encounters sowed discord, not only among the teachers but also among the students, each eager to defend his master.[87]

Maffei makes clear that what was at stake in the debate was not the best way to arrive at philosophical truth but the content and approach that were most appropriate and useful for training students as future scholars, perhaps, but also as members of a common religious and political community. It was true, Maffei argued, that his opinions could be countered by individuals who claimed that in this profession one must seek the truth, not what was useful. Such an enterprise was valid, he wrote, when one engaged in a philosophical dispute. Yet, he continued, university teaching was intrinsically tied to "public interests and political debates," and as a result, one needed to proceed differently. The goal was to determine not which doctrine was truer but which one was most becoming to the "tranquility of the State, the peacefulness of the people, and good political and civic governance."[88] It was also important that what was taught in the university was well regarded by the Church and contributed to its stability. Here Maffei praised the Jesuits, who were reputed for their judgment, their wise intentions, and their political savvy. According to Maffei, these Jesuit fathers "study, teach, and profess the Peripatetic doctrine," but only to the extent that it serves them to arrive at their own ends, namely, "to persuade others of what is beneficial to the Jesuits themselves." They teach Aristotle but "undermine [him] with interpretations and such comments" so as to make those of lesser understanding believe that Aristotle said what he did not. And, Maffei went on, if the Jesuit fathers, "men of perceptive intellect and of extraordinary reading and who continually apply themselves to the study of all arts and sciences," have understood that embracing Aristotle is necessary to achieve their goals, so, too, should those in charge of the Pisan university.[89] In short, it was not precisely the content and methods of the innovators that provoked concern; it was the way these individuals chose to transmit them in the university setting.

There are also suggestions that personal slights, not merely intellectual disagreement, fueled the bad feelings. In the midst of accusing the conservative faction of repeated and unprovoked hostility, Borelli added as the climax of his argument that the latest enmity was the result of the younger Bellini's oversight in failing to acknowledge Moniglia's work in his most recent publication: "But who isn't familiar with the very stoic patience of Bellini, who has tolerated so many insults over time that he has made himself an object of careful study without ever complaining, and just recently Mr. Moniglia named himself the prime enemy of the said Bellini for not having been thanked and praised with citations in his latest printed work; as a result he was obliged to receive the hits of and then kiss the [offending] cane."[90] Borelli's point was that his conservative colleagues were unreasonable in their expectations. His depiction, however, also makes clear that the conflict between innovators and traditionalists at Pisa was not purely cerebral. In this case, there was an art of courtesy governing the rules of citation, which Bellini had violated. The traditionalists on the faculty, who by and large had begun teaching at Pisa about a decade or more before the "Galilean" faction, expected acknowledgment and respect for their scholarship, for their training, and for their greater experience.[91] Their complaints derived from discomfort with the intellectual program of their younger colleagues, to be sure, but also from the feeling that they were being made obsolete before their time.

What this episode from Pisa, combined with the close attention to actual teaching there, suggests is that the apparent divide was a product of the rhetoric of the time and not intrinsic to scholarly practice. What brought individuals to use such rhetoric and to speak of fundamental differences, separating innovators from traditionalists, were specific historical, cultural, and social circumstances, in this instance antagonism and jockeying within a university faculty. In terms of their actual scholarly practices, both so-called innovators and traditionalists were much closer than they initially seem. Pisan professors, like Galileo himself and virtuosi readers like Wren, Ward, Viviani, and Mersenne, applied an eclectic mix of traditional and novel methods to the *Two New Sciences*. They used these techniques to incorporate the book into a variety of scholarly projects. While Pisan professors, on the whole, appear to have employed a more restricted range of approaches than did our earlier annotators, this difference likely reflects both their scholarly interests and the parameters of their teaching at Pisa.

CHAPTER SIX

Jesuit Bookish Practices Applied to the Two New Sciences

The *Two New Sciences* was a troublesome text. Its author, tainted with suspicions of heresy, proudly proclaimed his embrace of novel methodologies. The book itself did not fit easily into established genres, though it was read with methods and goals common to period scholarship. To make sense of the work, readers employed a variety of strategies, interpreting it as a source of mathematical problems, as a commentary on traditional natural philosophy, and as a repository of experiments. This chapter, which pairs with the previous one, considers how another local community, natural philosophers at the Jesuit Collegio Romano, responded to the difficulties posed by the *Two New Sciences*. Founded in 1551 by Ignatius Loyola (1491–1556), by the seventeenth century the Collegio Romano stood at the center of a vast international network of Jesuit colleges that aimed to provide a rigorous and orthodox education.[1] Similar to their contemporaries at Pisa, these philosophers responded in discipline-specific ways to the *Two New Sciences*. Their appropriation of the book for their own ends, however, went further as they domesticated the text using the bookish methods of traditional natural philosophy.

In 1658, twenty years after the publication of Galileo's *Two New Sciences*, Andreas Portner, a graduate of the Jesuit Collegio Germanico, shepherded into print the lecture notes of his professor of natural philosophy, Silvestro Mauro (1619–1687), then a professor at the Collegio Romano. These lecture notes formed the first teaching text produced by a philosophy professor at the Collegio Romano that cites Galileo's *Two New Sciences* specifically. Born in Spoleto, Mauro had studied philosophy and theology at the Collegio Romano from 1639 to 1648. After teaching first at the Jesuit school at Macerata and then at the Collegio Germanico in Rome, Mauro was appointed a professor at the Collegio Romano, where he taught theology, ethics, and sacred scripture in addition to the three-year philosophy sequence, comprising logic, natural philosophy, and metaphysics. Mauro remained at the Collegio

Romano until his death in 1687, serving both as prefect (1682–84) and rector (1684–87). In addition to his published lecture notes, which Portner had titled *Quaestiones Philosophicae*, Mauro also had printed a six-volume paraphrase of Aristotle's writings and two theological works.[2]

The *Quaestiones* is a compendium treating the standard Aristotelian texts taught during the Jesuits' three-year philosophical sequence. Following the Scholastic tradition, Mauro divided his volumes first according to the books from the Aristotelian corpus. The text itself is further subdivided into a running numeration of *quaestiones*. As traditionally employed in medieval university teaching, the *quaestio* followed a set format that served to organize discussion and commentary on given topics. After the question to be considered was presented, a list of arguments opposed to the preferred answer was given. These arguments were then refuted and resolved into the favored view using syllogistic reasoning and examples and definitions from commentators.[3] Mauro deviated from this set format slightly by dividing his response to the opposing arguments into two sections. After presenting the initial set of *sententiae*, or opinions, regarding the subject under discussion, he offered his own take on the question. He then turned to the opinions that opened the *quaestio* and evaluated each one in turn.

The production of a text designed to commemorate or facilitate one's classroom teaching was not unusual in the seventeenth-century Collegio Romano. Of the sixty-eight professors who taught philosophy at the Collegio between 1600 and 1700, at least seventeen produced manuscript treatises that are still extant in Roman libraries.[4] While many of these texts were composed by the professors themselves, others were copied down by students.[5] For the most part, these manuscripts were organized according to the curriculum specified in the 1599 *Ratio Studiorum*, a set of administrative rules intended to establish standards for the functioning of all Jesuit universities.[6] The *Ratio* declared that the philosophical sequence was to be taught as a three-year cycle as part of the upper-division curriculum. Logic was taught in the first year, while the second-year curriculum focused on Aristotle's *Physics*, his *On the Heavens*, the first book of his *On Generation*, and his *Meteorology*. In the third year, professors taught the remainder of *On Generation*, *On the Soul*, the *Metaphysics*, and the *Ethics*.

Other professors and students followed the lead of Portner and Mauro and published printed treatises on natural philosophy commemorating classroom teaching. Like Mauro's, these texts were designed as compendiums covering the writings of Aristotle laid out in the *Ratio*.[7] However, they were not all printed, as Mauro's *Quaestiones* was, as multivolume sets designed for easy portability. The 1625 *De Universa Philosophia* of Pietro Sforza Pallavicino (1607–1667) and Giovanni Battista Tolomei's (1653–1726) 1696 *Philosophia Mentis et Sensuum*, for example, are bulkier, one-volume, folio publications that may have been designed as reference tomes for professors preparing for lecture. In contrast, those texts published in the middle decades of the century, including Mauro's *Quaestiones*, were multivolume sets. The small size of these volumes, which range from vigesimo-quarto to octavo, suggests that they were intended to be portable and perhaps also more affordable.[8] The practice of teaching from shorter books was well established in the period. Sixteenth-century humanist teachers of ancient Greek had tended to choose small-format Greek-Latin editions of short texts, rather than great critical editions of Greek authors, for use in their classrooms.[9] Frans Titelmans's *Compendium Philosophiae Naturalis*, the bestselling handbook of natural philosophy first published in 1530 and used by Ulisse Aldrovandi (1522–1605) to prepare his lectures in Bologna, was similarly small in format.[10]

Opening Mauro's text, one finds his first mention of Galileo's *Two New Sciences* in a *quaestio* entitled "Whether some element is positively light." He attributed the first of his opening opinions to Galileo, whose character Salviati in Day 1 claims, contrary to established Aristotelian teaching, that no element has positive lightness:

> Here we see the true opinion of the Platonists and other ancient philosophers, whom Aristotle cites in book 4 of *On the Heavens* and in other passages, and to which Archimedes, like many other mathematicians, was partial, and which Galileo and other more recent philosophers follow: that positive lightness should be granted in no element or mixture, but all sublunar bodies are heavy. Granted, the less heavy are pushed upwards by heavier bodies, which descend with greater impulse and settle themselves beneath those that are less heavy.[11]

Mauro went on to reject this strong assertion and instead claimed that the element of fire was positively light.[12] After offering two proofs for the positive lightness of fire, Mauro conceded that it was nevertheless probable that other elements, including air, did not have positive lightness.[13] To bolster this assertion, Mauro offered a variety of proofs based on Aristotle's writings and common experiences.

Following his demonstration of air's positive heaviness, Mauro argued for an experimental measure of the heaviness of the air, which he obtained from the *Two New Sciences*. This reference was included in an aside that was distinct from the textual examples drawn from Aristotle and used in his proof. Mauro noted, "At this point, indeed, we may add a means [derived] from Galileo for weighing the air and for testing what proportion the heaviness of air has to the heaviness of water of the same volume."[14] In Day 1 of the *Two New Sciences*, Galileo put forward two different methods for determining the heaviness of air. Mauro described the second, which relied on one flask.[15] Mauro proposed filling a shallow dish with a very narrow opening with water and then weighing it. At this point, the valve would be opened and air allowed to exit freely. When the dish was weighed again, it would be lighter than before. The difference between the two measurements would correspond to the heaviness of the escaped air, the volume of which, Mauro asserted, was equal to the amount of water introduced into the dish. According to Mauro, Galileo relied on this method to determine that for the same quantity, air is nearly four hundred times less heavy than water.[16] Mauro endorsed the same conclusion as did Galileo for the results of the procedure: contrary to some interpretations of Aristotle, air did have heaviness, and its heaviness could be measured quantitatively.

Mauro also referenced Galileo's results on local motion in a *quaestio* that considered "whether and why the motion of heavy and light objects is accelerated." After establishing that such objects do undergo acceleration, Mauro asked his readers to consider the proportion by which this acceleration occurred. In his response, Mauro referenced the results Galileo presented in Day 3 of the *Two New Sciences*, answering that accelerated motion goes according to the odd numbers. As proof, Mauro cited Galileo's description of his experiments with the inclined plane.[17]

Mauro's approach to Aristotle and Galileo was in keeping with the aims and practices of Jesuit pedagogy. In the Jesuit curriculum, natural philosophy was considered an essential prerequisite to theology and superior to mathematics. It was organized around the teachings of Aristotle, whose writings were digested via a set of established questions drawn from the commentary tradition. The Society of Jesus required professors to promote uniformity and obedience to authority in their teaching while simultaneously offering a rigorous, up-to-date curriculum. These standards were enforced through a variety of censorship mechanisms, including internal systems of review, the promulgation of lists of banned opinions, and the development of general guidelines for correct theological and philosophical viewpoints.[18] Modern writers, even controversial ones, including Galileo, Copernicus, Descartes, and Gassendi, were thus folded into the curriculum, and their opinions were addressed when they touched on established questions.[19]

Mauro's work had a strong influence on the teaching of natural philosophy by Jesuits in Rome. A second, expanded edition of the work was published twelve years later, and some of Mauro's successors at the Collegio Romano likely culled parts of their own lectures from his publication.[20] This reuse of Mauro's text highlights the conservative nature of textbooks and teaching notes as scholarly productions. Such works—in the early modern period as today—were often reproduced and recycled by authors and their colleagues.

Mauro's citations of the *Two New Sciences* also set a precedent for subsequent professors' treatment of Galileo's findings. When they did mention the book in their teaching, these Jesuits focused, as had Mauro, on Galileo's odd-number rule and his measurement of the heaviness of the air. Jesuit discussions of Galileo's science of motion suggest that an internal commentary tradition was responsible for the propagation of a specific interpretation of Galileo's claims. How professors presented Galileo's measurement of the heaviness of the air reveals that bookish practices minimized the passage's novel elements, transforming Galileo's reliance on quantitative and experimental methods into mundane textual excerpts. Jesuits' decisions to teach Galileo should be interpreted less as a result of individual scholarly predilection for novelty and more the result of their scholarly practices.

An Internal Jesuit Commentary Tradition

The propagation of similar renderings of Galileo's findings on motion in Jesuit texts suggests that professors' evaluations of his science of motion were based on an established textual tradition within the order. This tradition was dedicated to commentary on and explanation of both ancient and new texts according to conventional natural philosophical topics and questions. It aimed not to provide a mathematical, experimental, or metaphysical validation for Galileo's rules—the goals of the "ideal" reader of chapter 1—but to comment on and relate them to set questions in the natural philosophical tradition. As did other readers, Jesuits largely read the *Dialogue* and the *Two New Sciences* independently and saw the two works as speaking to separate disciplinary conversations.

Mauro's inclusion of specific details suggests that he relied on the *Two New Sciences* when composing his discussions of Galileo's science of motion. To explore the proportion by which accelerated motion occurs, Mauro described Galileo's experiment with the inclined plane reported in Day 3 of the *Two New Sciences*: "Galileo affirms that he established by a very exact experiment that with a sort of canal having been elevated [and] divided into four equal parts, if a small lead ball is let loose on it, it traverses in equal time the first part of the canal and the three remaining [parts] as well."[21] Mauro continued by citing Galileo's results. In particular, according to Mauro, Galileo had found that "the velocity grows in equal times, according to the following odd numbers equally spaced 1, 3, 5, 7, 9, and so on, in such a way that if the heavy object in the first moment traverses a space of one palm-width, in the second equal moment it should traverse three palm-widths, in the third, five palm-widths, etc."[22] Mauro answered that accelerated motion goes according to the odd numbers, a conclusion Galileo presented as part of his discussion of naturally accelerated motion in the *Two New Sciences*.[23] Mauro's inclusion of specific details of the experiment with the inclined plane, reported in the *Two New Sciences* but not in the *Dialogue*, suggests that this passage drew either directly or indirectly on the *Two New Sciences*.

Despite his reliance on the *Two New Sciences* in this section of his teaching text, other passages reveal Mauro's familiarity with and willingness to teach his students about the prohibited *Dialogue*. He addressed Galileo's *Dialogue*, including passages treating local motion, in relation to Aristotle's *On the Heavens*. In his presentation of the world system, Mauro identified Galileo "in his dialogues published on this matter" as a follower of Copernicus, along with Kepler, Gilbert, Longomontanus, and others.[24] Later, Mauro addressed specific arguments from the *Dialogue*. He noted that Copernicans in general and Galileo in particular drew an analogy between a moving ship and a moving Earth, noting that "if a ball is projected upwards on a moving ship, it will fall back on the head of the one throwing it." Therefore, Mauro declared, Copernicans assumed that the same thing happened in the case of the moving ship. Mauro instructed his readers to ignore this claim, however, since other writers had affirmed that the experiment did not turn out as described.[25] He also rejected Galileo's argument based on the motion of the tides presented in Day 4 of the *Dialogue*. Referencing Niccolò Cabeo's commentary on Aristotle's *Meteorology*, Mauro noted that the tides followed the movement of the Moon and not the pattern claimed by Galilean Copernicans.[26]

Like Bérigard and Renieri at Pisa, Mauro read the *Dialogue* separately from the *Two New Sciences*. Whereas he taught the former in conjunction with the cosmological section of his natural philosophical course, he incorporated the latter into discussions of the terrestrial elements.

A closer look at Mauro's treatment of Galileo's findings on local motion reveals that while he was familiar with the *Two New Sciences*, his reading was not very accurate, for Mauro's restatement of the odd-number rule is incorrect. Whereas Galileo had asserted that the spaces through which an object accelerates in each instant of time are proportional to the odd numbers, Mauro claimed that it was the object's *velocity* that increased in this ratio.[27] Despite this error, Mauro did apply the rule correctly at the close of the excerpt. He noted that "if the heavy object in the first moment traverses a space of one palm-width, in the second equal moment it should traverse three palm-widths, in the third, five palm-widths, etc."

If Mauro's mistaken claim were an isolated occurrence, it would be tempting to dismiss it as the result of carelessness. However, later in his commentary Mauro erred again in his statement of Galileo's rule. In his attempt to explain the diminishing velocity of objects moved in violent motion, Mauro posited that these projectiles lose velocity according to the same proportion with which naturally accelerated objects augment it: "It is very reasonable that this effect of retarding acts according to that proportion by which the descent is brought about. But the descent possibly is brought about in such a proportion that if in the first moment the heavy body descends with a velocity of one, in the second equal moment it would descend with a velocity of 3, in the third with a velocity of 7, as is said by Galileo and others in Quaestio 49."[28] Again, Mauro stated that it was the velocities, not the spaces, whose proportions were augmented or diminished according to the odd numbers.

Mauro was not the only Jesuit professor at the Collegio Romano to present Galileo's definition of naturally accelerated motion in this way. Antonio Baldigiani (1647–1711), a professor who also served as consultor to the Congregation of the Index and was a respected member of scientific circles within and outside the Jesuit order, did the same in a 1696 manuscript teaching treatise on physicomathematics.[29] In this work, Baldigiani presented Galileo's findings on local motion in a section treating mathematical progressions or series. While the natural numbers were significant for their correspondence with all types of continuous quantities, including motion, time, and impetus, the series of the odd numbers beginning from unity was relevant for calculations involving naturally accelerated motion. According to Baldigiani, "Galileo in fact in his books on local motion demonstrates that heavy bodies descending naturally and freely within the same medium will increase their velocity and motion by single instants or by equal individual parts of time, so that *velocity or, what is the same, the spaces* run through by the movable grow in single instants by that arithmetic proportion by which the odd numbers increase in a natural series, namely, 1, 3, 5, 7, 9, 11." Baldigiani went on to repeat his statement that velocity and spaces are equivalent, noting that Galileo "rightly infers that velocities and motions or the spaces run through in such a descent" will be proportional to the squares of the times.[30]

Like Mauro, Baldigiani appears to have drawn on the *Two New Sciences* to address Galileo's science of motion, not the *Dialogue*. Baldigiani referred to Galileo's "books on local motion," a common expression period readers used to refer to Days 3 and 4 of the *Two New Sciences*. This phrase followed Galileo's own description of Days 3 and 4 as comprising a reading of "books on motion" written by his character the Academician. At the start of Day 3, the Academician characterizes his treatise as one divided into three parts, titled, in turn, *De motu locali* (On local motion), *De motu naturaliter accelerato* (On naturally accelerated motion), and *De motu proiectorum* (On the motion of projectiles).[31] In other sections of his treatise, Baldigiani addressed the subject of the tides, and he referenced explicitly the *Dialogue*, describing it as Galileo's "dialogues on the system of the world."[32]

Baldigiani also followed Mauro in erring in his statement of the odd-number rule but applying it correctly in practice. Instead of asserting, as Galileo had, that the spaces were proportional to the squares of the time, Baldigiani claimed that both the spaces and the velocities were proportional to the squares of the time. Yet, in later sections of the text he relied on Galileo's rule in its correct form to solve a number of problems. One of these asked readers to calculate how long it would take an object to fall from the Earth's surface to its center, assuming that the object was known to fall three geometric steps in one minute-second.[33] Galileo had solved a similar problem using his findings on local motion in the *Dialogue*, where he considered the fall of a cannonball from the Moon's orbit to the center of the Earth.[34] Such calculations of the fall of bodies to the Earth's center were fairly common in astronomical treatises; Galileo's discussion in the *Dialogue* was intended to correct a solution contained in the 1617 *Disquisitiones Mathematicae de Controversiis ac Novitatibus Astronomicis* authored by Johann Locher, a pupil of Galileo's rival Christoph Scheiner (1573–1650).[35] Baldigiani's application of

Galileo's odd-number rule to this calculation thus may signal that he at times read the *Two New Sciences* and the *Dialogue* in tandem, or it may simply be an indication of Baldigiani's familiarity with this long-established genre of astronomical problems.

This persistent misrepresentation of Galileo's rule among professors of natural philosophy at the Collegio Romano suggests that they may have been working within an internal textual tradition. The emphasis on note-taking in Jesuit pedagogical practice and the importance of shared texts—whether dictations of lecture notes, printed textbooks, notebooks, or loose note slips—meant that a particular reading introduced by one Jesuit author could be propagated in the notes and commentaries of his peers.[36] If Mauro and Baldigiani drew from the same set of sources, or if Mauro served as Baldigiani's source, this could explain how both authors made the same mistake in stating Galileo's definition of naturally accelerated motion. There are other instances of Jesuits producing commentaries on period self-described "novel" works. Marcus Hellyer, for instance, has argued that sections of Kaspar Schott's 1664 *Technica Curiosa* can be read as commentaries on Robert Boyle's experiments. In his text, Schott described every experiment in Boyle's *New Experiments Touching the Spring of Air* and added his own remarks to them. Unlike the pattern observed with Galileo's writings on motion, Schott's seventeenth-century Jesuit contemporaries and successors largely did not incorporate his commentary on Boyle in their teaching.[37]

André Semery's (1630–1717) philosophical textbook, first published in 1674, provides a hint regarding which text may have served as the source for Mauro's and Baldigiani's discussions of Galileo's science of motion.[38] Semery referenced an unspecified rule of fall in a *quaestio* that he titled "On the acceleration of motion." Semery began by evaluating various possible causes of the acceleration of motion. After attributing it to the addition of new parts of impetus, Semery added the following comment regarding the proportion with which objects are accelerated: "From this you can easily understand how impetus can be augmented repeatedly. . . . More recent philosophers generally follow this doctrine, and they usually say that motion is extended or that the velocity of motion is increased by the multiplication of impetus. But it is impossible to know by what proportion it is increased. . . . It is certain that it is increased by that proportion by which the impetus is multiplied, [but] by what proportion it is multiplied, guess, if you are able."[39] Semery's allusion to the confusion of philosophers was likely occasioned by the debate on Galileo's rules of fall that Paolo Galluzzi has termed the "second Galileo affair."[40] Because many fellow Jesuits contributed to the controversy, Semery may have believed it was safest not to specify a specific rule for his students.

Semery's language, however, suggests that his principal source for the debate was not Galileo. In Day 3 of the *Two New Sciences* Galileo defines naturally accelerated motion in terms of the addition of "equal moments of swiftness" (*aequalia celeritatis momenta*).[41] When he stated his rule of fall, the asserted relationship was always between "spaces" (*spatia*), "times" (*tempora*), and "degrees of speed" (*gradus velocitatis*).[42] In contrast, Semery framed the relationship in terms of "impetus," arguing that "the velocity of motion is increased by the multiplication of impetus."

Semery's source may have been another Jesuit who participated in the second Galileo affair, Cabeo. In his 1646 commentary on Aristotle's *Meteorology*, Cabeo also relied on the term *impetus*. He began by defining impetus, or "impressed power," as being produced from the very motion by which it was moved jointly with the mover itself.[43] This notion guided Cabeo's analysis of Galileo's results with inclined planes, pendulums, and projectile motion. Like Mauro and Baldigiani, Cabeo repeatedly erred in stating Galileo's findings on local motion. He claimed that it was the speeds, not the distances, that increased according to the odd numbers. For example, to facilitate his calculation of the ascent and descent of projectiles, Cabeo told readers he would assume that Galileo's definition of naturally accelerated motion was correct. According to Cabeo, he assumed as true "what Galileo said . . . that the velocity of a falling body increases in the duplicate proportion of the time, that is, as the square of the time."[44] In another section, Cabeo criticized Galileo's failure to provide a formal demonstration of his definition of naturally accelerated motion. Galileo's starting point, according to Cabeo, had been experimental. Citing the odd-number rule incorrectly, Cabeo noted that Galileo had moved from these experimental beginnings to "contend that velocity grows in equal times according to the odd numbers 1, 3, 5, 7, 9, and so on."[45] Despite these multiple misstatements of Galileo's findings, Cabeo, like Mauro and Baldigiani, did apply them correctly in sample problems that he included in his text.[46]

There is a good case for arguing that Cabeo's and subsequent readers' "mistakes" actually derive from incommensurable conceptions of velocity. Galileo's description of naturally accelerated motion relies on the notion of instantaneous velocity. In his scheme, the spaces traversed by a movable are in proportion to the squares of the times, and the velocities are in proportion to the times themselves. These proportions result in the odd-number rule, namely, that the space traversed in each instant of time increases as the odd numbers do. Cabeo, however, focused not on instantaneous velocities but on the average velocity across each interval. Taking the velocities to be uniform average speeds during each interval, these uniform velocities do increase as the spaces and thus as the odd numbers, which is what Cabeo claimed in his text.[47]

Whatever the underlying reason for Cabeo's misrepresentation—error or incommensurable conceptions of velocity—the repetition of the misstatement, combined with reliance on Cabeo's terminology, suggests that it was internal copying and a commentary tradition that led to its propagation. Several professors at the Collegio Romano referenced Cabeo's *Meteorology* directly, evidence that they had read and consulted either his work or an intermediate text that relied on him as they drew up their teaching texts.[48]

Further evidence of Cabeo's influence can be found in Giovanni Battista Riccioli's 1651 *Almagestum Novum*. Riccioli and Cabeo were both active participants in the Jesuit community at Parma and Bologna, a group known for its commitment to experimental and observational methods and interest in the latest developments in astronomy and natural philosophy.[49] Like Cabeo, Riccioli also erred in his statement of Galileo's odd-number rule, writing that Galileo had claimed that a movable's increase in velocity "was made according to the odd numbers counted from one."[50]

Mauro's and his colleagues' reliance on Cabeo to explicate Galileo's science of motion exposes the shared yet separate trajectories of the reception of Galileo's *Two New Sciences* and his *Dialogue*. In their teaching commentaries, Mauro and Baldigiani made clear distinctions in their treatments of Galileo's two publications, referring to the *Dialogue* when they discussed issues relevant to cosmology and to the *Two New Sciences* when dealing with questions of local motion or other philosophical topics found in Day 1. Cabeo, in contrast, purported to treat "Galileo's doctrine," but his discussion was based on Galileo's *Dialogue*.[51] Riccioli offered a comprehensive discussion of sources of and debates on midcentury astronomy, but he never cited the *Two New Sciences*.[52] By ignoring the 1638 text, Riccioli intimated that like many period readers, he judged Galileo's final published works as belonging to separate disciplinary spaces, even if they treated a common subject.

Jesuit readers thus approached Galileo's science of motion using the textual methods of commentary and note-taking associated with Aristotelian natural philosophy. Galileo's work, in this model of scholarship, was transmitted to Jesuit readers through their colleagues' commentaries. By responding to Galileo in this way, they signaled the continuity they saw between Galileo's approach and their own intellectual project of explicating Aristotle.

Domesticating a Troublesome Text

When Jesuit professors included the *Two New Sciences* in their curriculum, they introduced something new and potentially subversive. Galileo's text, even when it addressed standard Aristotelian questions, relied on novel methods, often to argue against received opinion. Jesuits who mentioned the *Two New Sciences* invariably inserted Galileo's experimental procedures and quantitative approach into their textual enterprise. Jesuit textual practices, however, worked to homogenize the problematic features of Galileo's book, minimizing its novel elements and emphasizing its links with Aristotelian philosophy. Jesuit treatment of Galileo's measurement of the heaviness of the air, reported in Day 1, is a good illustration of this phenomenon.

Galileo's discussion of his measurement of the heaviness of the air relies on elements of both traditional natural philosophy and developing experimental practice. Galileo employed commonplace examples and quotations from Aristotle's *On the Heavens* to put his findings in context. Alongside this more traditional discussion, Galileo described in detail two methods for determining the heaviness of the air compared with that of water, concluding that water is about four hundred times as heavy as air.[53] Jesuit readers who encountered this section of Galileo's *Two New Sciences* were drawn to it by its relevance to common questions addressed in contemporary natural philosophical texts regarding the heaviness of the elements and their motion. Aspects of Galileo's description, however, would have seemed unusual in this textual tradition, for Galileo appeared to provide specific details of an actual procedure and a quantitative result obtained by it.[54]

Jesuits managed these challenging aspects of Galileo's account by, first, excluding them from the demonstrative portions of their texts and, later, casting Galileo's account not as a record of experimental activity but as a report by a textual authority. In his *Quaestiones Philosophicae* Mauro relied on the former strategy. He argued that while fire had the quality of lightness, it was probable that other elements, including air, did not have positive lightness.[55] To establish that air was positively heavy, he offered a variety of proofs based on Aristotle's writings and commonplace experiences. An examination of Mauro's word choice reveals his conception that his argument—the heaviness of the air—was proved via an appeal to generalized experience, textual authority, and logical argument:

> The second part of the conclusion, that air has positive heaviness, *is proved* . . . because leather bags, *as Aristotle remarks in book 4 of* On the Heavens, *text 30,* weigh more inflated than when they are not inflated; therefore the enclosed air in the leather bags adds weight; therefore it is weighty and heavy. Adversaries can respond that our air is heavy because it is vaporous and contains many aqueous parts. However, since we find *by experience* that our air adds heaviness, since *other experiences agree,* and since we are not able to acquire any experience of that pure air, we ought to say that air is absolutely heavy, even if it is light with respect to earth and to water.[56]

In this passage, Mauro referred to textual authority—Aristotle's *On the Heavens*—and appealed to general experience. No reference was made to a particular experimental procedure, nor, aside from Aristotle's, is a specific experience mentioned. It was Aristotle's text, affirmed by Mauro's and others' "experiences," that allowed Mauro to claim that the air was heavy.

Following his demonstration of air's positive heaviness, Mauro departed from this purely textual debate. He followed his demonstration with a reference to Galileo's procedure: "At this point, indeed, we may add a means [derived] from Galileo for weighing the air and for testing what proportion the heaviness of air has to the heaviness of water of the same volume." "Take a shallow dish made of glass with a very narrow mouth, and a continuous shell or outer covering," he began, and he went on to describe in detail a procedure that closely resembled the one Galileo had included in his *Two New Sciences*. Mauro closed his description by noting, "By this calculation, Galileo held that he discovered that air weighs around four hundred parts less than does water of equal quantity."[57]

Mauro's text emphasizes the experimental and quantitative nature of Galileo's passage even as it minimizes their relevance to his demonstration. While Mauro agreed with Galileo that the air was heavy, he did not incorporate Galileo's measurement in his proof. For his demonstration, Mauro instead relied on a logical proof filled with textual references and universal experiences commonly reported in other texts. As the report of an experiment, Galileo's procedure was an interesting aside, not suitable for inclusion in a demonstration undergirded by textual evidence.[58] This impression is supported further by the asymmetry underlying Mauro's proof and presentation of Galileo's result. In the course of his proof, Mauro responded to the potential objection against Aristotle's example of the heavy leather pouches filled with air. According to this argument, the pouches were heavy not because air was heavy but because they were filled with a mixture of pure air and earthly vapors. Though this argument was applicable to Galileo's procedure, Mauro did not address the relationship between it and the counterargument. Galileo's reported experience and the generalized experience and logical arguments of textual authorities remained separate in Mauro's account.

In subsequent decades, professors began to integrate Galileo's experimental results more fully into their demonstrations. They were led to do so, however, less by an increasing acceptance of experimental evidence and more because they began to portray Galileo's procedure as a piece of bookish evidence that could be cited and discussed as a textual authority.[59] One indication of such a transformation is a set of manuscript teaching notes that follow a philosophical course offered in 1660 by Ignatius Tellin (1623–1699). Born in Armagh, a city in present-day Northern Ireland, Tellin joined the Jesuit order in 1642 and taught philosophy, mathematics, and theology in Venice and in Rome.[60]

Like Mauro, Tellin claimed that fire was positively light and that the other elements and mixtures were heavy. However, in place of a certain demonstration of the air's heaviness, Tellin offered an argument for its probable heaviness. In Tellin's words, "It is probable that they [the other elements and mixtures, aside from fire] do not have positive lightness, but only less heaviness." Tellin then went on to list evidence in support of his assertion. For one, he noted, "bodies never rise above those heavier in species except when an impetus is impressed on them by a heavier one, forcing them out." As an illustration of this principle, Tellin pointed to another instance of generalized experience reported in Aristotle's *On the Heavens*, namely, the behavior of a pouch filled with air submerged in water. According to Aristotle, such a pouch will attempt to ascend, but when the water is frozen, the pouch is forced to remain under the frozen water unless it manages to break the ice. Such a scenario, argued Tellin, was comparable to the behavior of air trapped in stones and metals, which would also be inclined to ascend. More evidence, claimed Tellin, supported his contention of the air's heaviness. Tellin pointed, first, to the example noted by Mauro from Aristotle's *On the Heavens* of the inflated leather pouches. The increased heaviness of the pouches when inflated, according to Tellin, "is evidence . . . that simple, unmixed air is heavy and weighty." Such a conclusion, argued Tellin, was analogous to Galileo's recognition "that the heaviness of water to the heaviness of air is in the proportion of 400 to 1."[61]

Although Tellin reached the same conclusion as Mauro, the role he allotted to Galileo's measurement was very different. For one, he included Galileo's procedure as evidence in support of his claim of the air's heaviness, whereas Mauro had relegated Galileo's measurement to the status of an addendum. In addition, Tellin's word choice portrayed Galileo's recently reported measurement to be on equal footing with the commonplace examples drawn from Aristotle. Tellin divided his four examples into pairs, which he connected by the Latin words *his accedit*, or "to these it is added." He then equated the two members of each pair by the Latin words *sicut* (just as, in the same way as) and *ita ut* (in the same way as). The behaviors noted by Aristotle of the leather pouches were equivalent, in Tellin's writing, to the general observation of air trapped in stones and Galileo's specific measurement of the heaviness of air compared with that of water.

Galileo's procedure as relayed by Tellin, however, may not have been conceived as a reported experiment at all, but merely a passage describing an experiential observation. Unlike Mauro, he did not describe Galileo's experimental procedure but only referred to Galileo's "recognition" of the quantitative measurement. Tellin may have been led to render the passage in this way because he read about it in Mauro's text; his word choice and sentence structure indicate that he likely developed his course using Mauro's printed textbook as a model.[62] If so, perhaps Tellin never consulted the *Two New Sciences* directly but judged Galileo's claims reliable because they were reported by Mauro. If that were the case, Tellin would have viewed and written about Galileo's claims as textual excerpts to be incorporated into his demonstration because he had encountered them as such in the published book of his colleague.

Later professors likely did not use Mauro's text as a model, but they did portray Galileo's empirical evidence as a textual claim, not the report of an experiment. Unlike Mauro, who described in detail Galileo's procedure, these professors limited themselves to citing Galileo's claim to have experimented or to stating only Galileo's numerical measurement. For example, Ottavio Cattaneo, in his 1677 *Cursus Philosophicus*, argued that air, like fire, was light and not heavy. In support of his conclusion, he addressed Aristotle's example of the inflated pouch weighing more when full than when empty. According to Cattaneo, this observation could be explained by the fact that "the ball receives weight from the thickness violently impressed on it when it is inflated by hand." Furthermore, continued Cattaneo, this same explanation could account for Galileo's own measurement of the heaviness of the air. In his words, "This is also the reason why sometimes we find that air that strikes the head is thick and heavy, since without a doubt it is mixed with many impurities, which perhaps provided an opportunity for Galileo, as he claims, to weigh the air. For breathable air is neither clear, nor is it the most pure, since it is not entirely purged of all defects from foreign bodies."[63] Cattaneo transformed Galileo's report of his experiment into a textual claim, one that could be countered by a standard philosophical argument. Cattaneo argued that both the air in the inflated leather pouches and the air weighed by Galileo were not pure but merely breathable air corrupted by terrestrial and aqueous elements.

Though some Jesuit authors at the end of the seventeenth century appear to have privileged quantitative, experimental findings in their teachings, they too emphasized the origin of these claims in texts, not in experimental trials. As a first example, consider Tolomei's 1696 *Philosophia Mentis et Sensuum*. In his discussion of the terrestrial elements, Tolomei addressed the properties of air. He organized this part of his course as a series of conclusions that he stated and then proved. Under the conclusion, "Air has absolute heaviness," Tolomei began not by considering Aristotle's text but with a declaration of the findings of recent writers: "I pronounce at the start that air is respectively lighter than water, or less heavy in the proportion, according to Galileo, as 1 to 500; according to Mersenne and Fabri, etc., as 1 to 1,000."[64] Tolomei cited these experimental findings, but as in traditional philosophical demonstration, he emphasized their origin in the writings of other authors. The question for students was not what the precise ratio of the heaviness of air to water was but whose textual claims they should believe, Galileo's, Mersenne's, or Fabri's. Tolomei's continuing reliance on textual methods is revealed in his demonstration of the air's heaviness. In this section Tolomei based his conclusions on traditional, textual arguments. He cited various passages from Aristotle's *On the Heavens*, as well as common textual examples, including that of the inflated leather pouch, in the course of his demonstration.[65]

Similar ambiguities are observed in the manuscript teaching notes of Giovanni Iacopo Panici (1657–1716), who taught natural philosophy at the

Collegio Romano from 1698 to 1701.[66] Panici began his discussion of the air's heaviness by declaring that "the heaviness of the air cannot be explained more appropriately than by comparison with the heaviness of other bodies, whose heaviness is perceived by the senses. If air therefore would be compared with water through a ratio of heaviness, it will have [the ratio] of 1 to 500, if we believe Galileo; or if we believe Mersenne, Fabri, and others, of 1 to 1,000."[67] Panici thus addressed the traditional question of the heaviness of the air in a way that resembles Tolomei's, by beginning with recent experimental, quantitative measurements of the heaviness of the air. The air's heaviness could not be explained "more appropriately," stated Panici, than by comparing it quantitatively with the heaviness of other bodies. While Panici later did cite Aristotle's *On the Heavens* as confirmation of his supposition that air was heavy, he at no point provided a traditional, Scholastic proof of its heaviness.[68]

These features may suggest that Panici abandoned the bookish tools of traditional natural philosophy in favor of new experimental evidence. Yet, such a conclusion would be too hasty. For one, even in this initial statement Panici emphasized the values reported by Galileo and others, but he did not emphasize the act of experimentation. Panici also revealed his bookish methods even when he adopted some of the rhetorical tactics of well-known proponents of the New Science. In his discussion of the recent experiments with the Torricellian void, Panici criticized the Aristotelian notion that nature's horror of the void underlay the behavior of the mercury. He argued that if nature's abhorrence of the void were responsible for the mercury's behavior, this virtue or quality of abhorrence would have to be extremely clever. To account for all observed behaviors, it would need to distinguish between long and short, inclined and perpendicular tubes and be able to recognize when the tubes were placed in valleys or mountains.[69] All of these claims resemble the standard rhetoric used by detractors of Aristotle, including Galileo himself.

It was when Panici turned to the common technique of appealing to numerical results that he revealed his adherence to textual sources. Galileo offered many models of how to use numbers to refute textual claims. Consider, for example, the section of Day 1 of the *Two New Sciences* in which Salviati urges Sagredo and Simplicio to reflect on their personal experience with falling objects. Though Aristotle claimed that a one-hundred-pound iron ball falling from a height of one hundred braccia should hit the ground before a ball of one pound had descended a single braccio, Salviati assured his listeners that their own experience would show them that the smaller ball would only lag behind the larger by two inches.[70] These numerical, immediate, and personal experiences were central in Galileo's text for demonstrating that Aristotle's statements were incorrect.

Panici did something similar in this section of his treatise. According to Panici, there was an additional reason why the behavior of the mercury in the tubes could not be explained by the Aristotelian notion of the horror of the void. It was that such a supposition left many phenomena unexplained, in particular the natural limit that had been observed in the ascension of bodies. Aristotle had stated that bodies always move to fill a potential void because such a void is prohibited by nature. Panici argued that this explanation must necessitate that bodies be able to rise without limit to fill such potential voids. Yet, noted Panici, referring to a passage in Day 1 of the *Two New Sciences*, Galileo himself "discovered that this was obviously false, and it is clear from everyday experience, by which it is proven that water can only ascend in pipes up to a height of about eighteen cubits."[71]

At first glance, it appears that Panici employed the claim for the water's maximum height in much the same way as Galileo had his numerical example. According to Panici, since Aristotle's explanation does not account for the fact that liquids have a maximum and particular height to which they can be raised, Aristotle's line of reasoning cannot hold. Closer examination, however, reveals a subtler distinction between Panici's and Galileo's citation of these examples. Whereas Galileo, through Salviati, refuted Aristotle based on personal experience, Panici claimed that the recent *Two New Sciences* contained measurements with truth claims superior to those found in Aristotle. In Panici's account, a recent text containing reports of experiences—not the experiences themselves—trumped the descriptive, textual claims of Aristotle.

Panici's text, like those of his predecessors, thus reveals how Jesuit professors reconfigured the novel aspects of Galileo's text. In the teaching of these professors, Galileo's claim was inserted alongside the arguments of traditional commentators by turning it into another textual example to be cited and discussed. Galileo's reported experiments and quantitative measurements did not disrupt the bookish practice of natural philosophy when they were treated as instances of it.

Why Teach Galileo?

Jesuit professors' treatment of the *Two New Sciences*, like Bérigard's at Pisa, indicates that their reading and teaching of the book was selective, focusing on only a few of its passages. An important question to consider is why these professors even made the effort to include the *Two New Sciences* in their classroom teaching at all. We might suspect that they included the book because it treated relatively safe topics that, if discussed carefully, would keep professors from engaging with Galileo's support of Copernicanism. The *Two New Sciences*, according to this interpretation, may have served as a vehicle for professors to address the famous Galileo with groups of young students eager to learn more.

A broader examination of period teaching at the Collegio Romano, however, reveals that this explanation cannot hold. Instead, Galileo's astronomical work, his forbidden *Dialogue*, and his condemnation received much more attention in teaching at the Collegio Romano than did his *Two New Sciences*.[72] Classroom discussion of Galileo reveals that he occupied a prominent place in the curriculum in sections dealing with cosmology or astronomy. His telescopic discoveries often merited special consideration. In an undated treatise on practical geometry, for example, the professor of mathematics Francesco Febei, who taught from 1665 to 1689, singled out Galileo two times as the inventor of the telescope in discussions of optics and the use of the telescope and the microscope. According to Febei, although "chance" had been responsible for the initial invention of the telescope, "its polishing and perfection are owed to Galileo, who improved it from its undeveloped state to such a degree that he is held by many as the Author and inventor of the Telescope."[73] In his 1646 commentary on Aristotle's *On the Heavens*, the professor of philosophy Ludovico Bompiani described Jupiter's satellites and emphasized Galileo's role as the first person to observe them.[74]

While Galileo was portrayed more negatively in discussions of the world systems, there too he was mentioned frequently and with some detail. There is no indication that Jesuit professors felt they had to hold themselves back from discussing his work, though they were required to pronounce against the Copernican hypothesis. In these sections, professors emphasized Galileo's status as a supporter of the condemned Copernican system. Mauro, for example, identified Galileo as a follower of Copernicus, along with Kepler, Gilbert, Longomontanus, and others.[75] Mauro's successors consistently named Galileo as a proponent of the famous, if mistaken, Copernican system.[76] Professors of mathematics went further, citing Galileo as an exemplum of what comes to pass when one disobeys the Church's injunction. In his 1689 *Cursus Physicomathematicus*, Francesco Eschinardi (1623–1703) told readers that Galileo had argued that the Copernican hypothesis was real, not a mathematical hypothesis, for which he had subsequently been condemned.[77] Many of these professors, like Mauro, explicitly cited and discussed specific arguments Galileo put forward in the *Dialogue* in sections of their commentaries treating Aristotle's *On the Heavens*.[78]

Jesuit professors thus did not turn to the *Two New Sciences* out of a frustrated desire to discuss the more controversial topic of Copernicanism. If they wanted to teach students about Galileo's *Dialogue* and condemnation, they could, and they did. Nor did they bring up the *Two New Sciences* because of its relationship to the more controversial *Dialogue*. By and large, these professors did not teach the two works in tandem. They saw the *Dialogue* as relevant to the discussions in *On the Heavens*, whereas the *Two New Sciences* was read in connection with specific questions, including the nature of the void, the heaviness of the air, and the nature of local motion.

Historians of the Jesuit order and early modern universities have often tried to answer the question, why Galileo? in relation not to scholarly practices but to scholarly predilections. The reception of novel texts like Galileo's *Two New Sciences* has been interpreted as an indication of individuals' or institutions' predisposition to innovative ideas. Such an agenda has led scholars to reevaluate the role of the Jesuit order and universities more generally in effecting transformations in seventeenth-century science.[79] It also has allowed for new perspectives on the mechanisms and efficacies of established censorship procedures.[80]

In this spirit, we could ask whether these professors' responses to the *Two New Sciences* reflect a rejection of the New Science or an openness toward it. The evidence is ambiguous. On the one hand, the narrative above could be interpreted as an indication that these professors engaged in such a limited way with Galileo's book that they clearly were traditional, conservative thinkers unimportant to a larger narrative of the transformation of early modern science.[81] Teaching notes and textbooks exist for fourteen of the more than forty-five professors listed as having taught the philosophy sequence from 1638 to 1700. Only six of these documents mention Galileo's measurement of the heaviness of the air, while only four discuss the possibility of determining a quantitative measure of the velocity of an accelerated body, with three of these citing Galileo's findings specifically. Focusing on the evidence of nonengagement, we would note the very late date at which Jesuit professors began incorporating Galileo's *Two New Sciences* in their teaching, the small number of individuals who did so, and their tendency to read Galileo through the lens of Aristotle as evidence that few professors were interested in Galileo and that of those who were, most did not completely understand or accept the novelty of his approach. There is, however, another way to read the sources. In the teaching texts of these relatively unknown Jesuit professors of philosophy, it is possible to locate, not a slavish adherence to tradition, but rather an engagement, albeit limited, with elements of the New Science and an effort to disseminate it to students.[82]

These two interpretations both have merit and precedence in Jesuit scholarship. Yet as their contradictory nature suggests, there is something unsatisfactory about framing our studies in this way. In the case of Jesuit pedagogy, this focus on scholars' penchants for novelty leads us astray, because the purpose of these teaching texts was not to embrace novelty or suggest an affinity with Galileo or other recent writers but to introduce the questions associated with the Aristotelian corpus to students and prepare them to defend established opinions with vigor. The teaching of philosophy at the Collegio Romano had developed from medieval Scholastic models. Originally intended as a technique for defending orthodoxy, the Scholastic method demonstrated truth by proving the falsity of opposing opinions. Following Aristotle's dialectic and his discussions of methods, students were instructed that the best way to arrive at the truth and persuade listeners of it was to examine arguments for and against given propositions. The university exercise of the disputation, which formed a public part of university education and in which both students and professors participated, was the perfect model of this technique.[83] Disputation was considered a central aspect of Jesuit pedagogy, with the ideal cycle of learning being *lectio—repetitio—disputatio*. Following lecture and recapitulation exercises, students were to exhibit proficiency in the material taught by calling it forth spontaneously and to demonstrate their own mastery of the subject.[84] Teaching texts prepared students for these exercises by posing questions, offering potential responses, and refuting incorrect ideas in order to prove more forcefully the rationale underlying the accepted answer.

A more convincing explanation for Jesuit interest in the *Two New Sciences* thus can be found in the eclectic practices and goals of traditional natural philosophy. At the Collegio Romano both printed and manuscript teaching texts in philosophy were expected to comprise a "reading" of Aristotle's writings on the natural world, but Aristotle's words received little attention in the notes themselves. Instead, the notes and the textbooks were made up of the questions that had accumulated around the Aristotelian corpus since late antiquity. Aristotle's discussion of the void, for example, would be reduced to a number of key queries, such as, what is a void? what types of void space are possible? and can motion occur in the void?[85] Historians have shown that in Jesuit colleges and elsewhere the topics and questions considered in such courses remained relatively static across time, though the specifics of the answers and the evidence amassed in response to them was more dynamic and more responsive to new speculation.[86] This openness was a key feature Schmitt identified when he labeled early modern Aristotelianism as an eclectic philosophy.

Reaction at the Collegio Romano to Galileo's speculations indeed resembles early responses by Jesuits more generally to the air pump. In their academic writings and their teaching treatises, Jesuits initially regarded the pump as simply another one of the many *experientiae* that demonstrated the impossibility of the void and were cited in the context of Aristotle's treatment of the topic in his *Physics*. Even when Kaspar Schott (1608–1666), the Jesuit priest who published the first account of the device, revised his understanding of it, he reinterpreted it as an experiment that revealed something about the heaviness of air, another question central to Aristotelian philosophy.[87] Aristotle's writings and the commentary tradition surrounding them thus provided the framework within which these novel devices and writings were interpreted.

When professors included, or neglected to include, Galileo's *Two New Sciences* in their teaching, then, they did so not necessarily as a testament to their predilection for or resistance to novelty or tradition but because they believed or did not believe that his arguments were pertinent to their scholarly project of explicating Aristotelian natural philosophy. While discussion of novel ideas may denote an affinity toward the new, it is better interpreted as an indication of the degree to which a given professor kept abreast of recent speculation and the connections he saw, if any, between his scholarly project—one oriented around Aristotle—and what he read. They are a signal of how these professors understood the relationship between their intellectual work and the New Science.

That professors' citations of Galileo reflect judgments of relevance, not attraction to novelty, can be seen in the pattern in which they incorporated Galileo's measurement of the heaviness of the air in their teaching materials. Most professors followed Mauro's lead and referred to the passage in the sections of their commentaries that treated the four terrestrial elements, earth, water, air, and fire.[88] There was good reason for these professors to believe that Galileo's procedure and reported measurement were directly pertinent to this part of the curriculum. Aristotle had discussed the four elements in many of his writings, most notably in his *On the Heavens* and *On Generation*. Jesuit professors drew this material together in sections of their curriculum that they titled "On the Elements." Here they treated questions related to the elements' qualities and motions. Students were required to identify the four elements, their qualities, and their motive virtues and be able to explain the role of heaviness and lightness in causing their motion.[89] Galileo's discussion of air's heaviness was thus directly related to these subjects.

Galileo's Jesuit readers were also interested in this passage because the question of the heaviness of the air had a long history in medieval Scholasticism and even antiquity. Galileo, who himself was educated in natural philosophy at the University of Pisa and was a serious student of Aristotle in his younger days, was familiar with this tradition and wrote the *Two New Sciences* with an eye toward it in some instances.[90] In his description of his two methods for measuring the heaviness of the air, Galileo referenced this tradition specifically through the comments of his interlocutor Simplicio, who cites Aristotle's *On the Heavens* directly, calling attention to Aristotle's example of leather pouches, which are found to be heavier when inflated with air than when they are empty.[91]

This example of the inflated pouches or bags was commonly cited by Aristotle's ancient, medieval, and early modern commentators. Aristotle's ancient commentator Simplicius treated the example at length in his own commentary on Aristotle's *On the Heavens*. Later medieval commentators, including Peter of Auvergne (d. 1304), drew on Simplicius's writings in their discussions of the heaviness and lightness of the elements.[92] In his commentary on book 4 of *On the Heavens*, for example, Jean Buridan (ca. 1300–after 1385) posed the *quaestio* "whether air is heavy or light in its own region, or whether it is neither heavy nor light." Buridan described the example of the inflated pouches twice in his discussion, both times as an example of Aristotle's conclusion that air does have heaviness in its region.[93] Because Buridan argued that air does not have heaviness or lightness in its region, he accounted for Aristotle's example of the pouches by appealing to the commentary of Ibn Rushd (Averroës) (1126–1198), who had argued that the air in the inflated pouches was heavier only because it was more condensed than the outside air.[94] Albert of Saxony (ca. 1316–1390) offered a similar presentation of the question and example. He discussed the pouches in a *quaestio* entitled "Whether some element in its place is heavy." Like Buridan, he cited the example as support for the claim that air has heaviness in its own region but dismissed the evidence by appealing to the explanation that its apparent heaviness is only due to its being condensed and compressed within the pouch.[95]

The example of the pouches was also singled out for discussion in the commentary produced by the Jesuit community at Coimbra, which was published in multiple editions and often cited in textbooks and teaching notes of professors at the Collegio Romano. The Coimbra commentary included Aristotle's Greek text alongside a Latin translation, followed by an explanation and various *quaestiones* pertinent to the passage in question. It was followed by a treatise on various problems relating to the four elements. The second of the problems concerning the element air elaborated on the difficulties posed by the pouches. It asked the question, "If pouches filled with wet air are heavier than empty ones, as we admitted above, why do those float on water, [while] the others sink?"[96] The commentators responded by ignoring the behavior of the empty pouches and arguing that the lightness of the air contained within the pouches filled with the "aqueous water" made the said filled pouches lighter than water, especially given the tendency of the light air to impel itself above water.[97]

In deciding to cite Galileo's *Two New Sciences* in their teaching, then, Mauro and his successors were following in the best of the Scholastic *quaestio* tradition. The objective of this enterprise was to explain and prove the correct responses to the individual *quaestiones* posed in response to an authoritative text, in this case Aristotle. Conclusive demonstrations rested on the ability of the writer or speaker to address all potential objections to the correct answer and show them to be false. Galileo's measurement of the heaviness of the air was just such an argument that professors wanted to include in their arsenal. Galileo presented it as closely related to topics traditionally associated with sections of Aristotle's *On Generation* and *On the Heavens*. Furthermore, the question of the lightness and heaviness of the elements was a controversial and much-discussed one among Renaissance Aristotelians, who addressed it in university disputations, in special treatises, and even through experiments.[98] Galileo, moreover, directly alerted readers to the connections between his text and this tradition by citing Aristotle's *On the Heavens* and the well-known example of the inflated pouches.

Discussion of Galileo in these teaching treatises is thus a sign that these professors were familiar with Galileo's writings, saw connections between them and their expositions of Aristotle, and were diligent scholars, interested

in providing comprehensive treatment of standard questions by incorporation of more recent material. The reception of the *Two New Sciences* in the
Collegio Romano was predicated on discipline-specific reading in which
professors sought in Galileo ties to Aristotelian natural philosophy. It was
not a referendum on Galileo's condemnation or a reflection of his status as a
famous astronomer or defender of the world systems.

The insertion of the *Two New Sciences* into classroom teaching of philosophy reveals why and how a group of readers committed to traditional scholarship read and refashioned the book to make it appropriate for their students. Galileo's gestures toward standard philosophical topics encouraged
discussion of it, especially because Jesuit teachers were enjoined to maintain
high standards of scholarship. The techniques they applied to read and write
about the *Two New Sciences* were similar to those used by readers at Pisa and
elsewhere. Jesuit professors read the *Two New Sciences* for the sections that
touched on their expositions of Aristotle and with the scholarly methods of
their discipline, largely ignoring the book's mathematical sections and proofs.
The textual nature of their own scholarly practice, in turn, provided a ready
mechanism for the domestication of the troublesome text, as the problematic elements of the *Two New Sciences*—its reliance on experimentation and
quantification—were transformed in these textbooks into textual claims to
be put in dialogue with established authorities.

Epilogue

In the previous chapters I have showcased readers who incorporated Galileo's *Two New Sciences* into diverse projects, including the accumulation of commonplaces, expositions of Aristotle, the perfection of mathematical techniques, and the stockpiling of ideas for new experiments. These approaches to the text represent a different type of reading than our own, one more attuned to the bookish methods of early modern Europe than the approaches of modern science. They contrast with Galileo's own repeated instructions to and criticisms of his contemporaries, exhorting them to engage in reading that verified the printed page through examination of nature. They also challenge existing scholarship on the text, much of which implies that historical actors read to verify Galileo's ideas. These differences are of significance. Seventeenth-century readers, despite Galileo's appeals and at times their own predilection for novel approaches, had not yet imbibed the reading methods advocated by Galileo and later embraced by modern science. Galileo's book, a text so beloved and ostensibly representative of new mathematical and experimental currents, was understood by seventeenth-century readers as a work in conversation with the bookish approaches its author professed to reject.

This last point is analogous to one made by Owen Gingerich regarding his census of extant copies of Copernicus's *De Revolutionibus*. Gingerich titled his account of his thirty-year odyssey to examine all extant copies *The Book Nobody Read*. The "never read" of the title referred explicitly to historians' prior claim that no one really took Copernicus's work seriously from its publication in 1543 until Galileo's telescopic discoveries announced in 1610. Gingerich, in contrast, found that professional astronomers, teachers, and students had pored over Copernicus's work, using the margins in the book to work through the text. However, Gingerich's title can also be understood in another way: Copernicus's *De Revolutionibus* was a book "nobody read" because the types of readings Gingerich uncovered were unlike those carried out after Galileo's and those now expected by modern historians. Whereas Galileo and increasingly others in the seventeenth century believed in the reality of the heliocentric hypothesis and sought to demonstrate correspondence between the physical world and Copernicus's mathematics, early readers by and large read Copernicus as a mathematical, predictive astronomer. They used his mathematical hypotheses to improve the accuracy of their own predictive models, not to challenge the accepted Aristotelian world-view.

This distinction between different purposes of reading is one addressed repeatedly in the previous chapters. Readers, both those who annotated the printed text directly and those who taught or took notes on it for their own scholarship or teaching, applied to the *Two New Sciences* a variety of reading methods drawn from the mixed-mathematical and natural philosophical traditions. They often did not read Galileo in the way he instructed his own readers to do so. Many of Galileo's readers, like Copernicus's own, were less interested in the relationship between his text and the real world than historians have assumed. They did not conform to the "ideal" reader expected in the historiography, one who read the *Two New Sciences* in tandem with the *Dialogue*, whose interest was in the physical veracity of Galileo's mathematical claims, and who sought to verify and extend Galileo's findings.

One of the main reasons historians have expected Galileo's readers to have read his work critically using new methods was because of Galileo himself. Beginning with the publication of his *Starry Messenger* in 1610, Galileo took care—through the letters he wrote, the works he published, and the attention he paid to the preservation of his papers—to portray himself as the instigator of a new way of studying nature. He emphasized the innovativeness of his own approach, contrasting the quantitative and experimental methods he embraced with the textual ones of his contemporaries. He was also keen to gain recognition for the novelty of his contributions. Period letters reveal that one of the findings he recognized as important was his science of motion presented in the *Two New Sciences*.[1]

While Galileo was instrumental in shaping his image and his legacy, the person most immediately responsible for promoting Galileo's vision was Viviani. Viviani pledged his life to securing Galileo's reputation not only as an accomplished natural philosopher, mathematician, and astronomer but also as a pious Catholic. He dedicated himself to a number of projects to achieve this goal: the reprinting of all of Galileo's works in Italian and Latin, the composition of a biography of his teacher, and the construction of a celebratory tomb alongside those of Florence's other great sons, including Michelangelo.[2] Viviani's labors are a testament to the persuasive force of his teacher. They also reflect a wider period trend for memorializing enterprises, from Pierre Bayle's (1647–1706) and Jean Le Clerc's (1657–1736) accounts of Erasmus to the collaboration that produced the 1623 First Folio of Shakespeare's plays.[3]

Viviani's work led to the accumulation of a wealth of Galilean memorabilia, as correspondents sent to him all of Galileo's letters and manuscripts in their possession. Because Viviani was committed to restoring Galileo's reputation as a pious natural philosopher, he kept abreast of his contemporaries' publishing activities and the Galilean materials they possessed. He urged his contacts to inspect surviving materials personally and to take whatever means necessary to ensure that potentially damaging materials would not be brought to light, lest they tarnish Galileo's already fragile reputation.[4] The surviving materials are thus a selective record of what Viviani wanted to preserve for posterity.

Viviani also sought to promote the aspects of Galileo's intellectual work that Galileo himself had touted as his contributions to contemporary science. The scholarly projects to which Viviani dedicated himself, including his 1674 *Fifth Book of Euclid's Elements*, were intended as extensions and elaborations of Galileo's *Two New Sciences*. The anecdotes Viviani recorded in his biography of Galileo, from his teacher's observations of the swing of the chandelier in the Pisan cathedral to his dropping of heavy objects off the leaning tower of Pisa, are now generally considered apocryphal.[5] They point, however, to Viviani's own desire to celebrate Galileo's science of motion, his quantitative approach, and his reliance on experimentation, even as Viviani's own reading and writing about Galileo testify to the more traditional ways he interacted with his teacher's scholarship.

Viviani's impulse to promote these aspects of Galileo's intellectual project undoubtedly shaped his choice of materials for collection. This emphasis was strengthened when the manuscripts were organized in the second decade of the nineteenth century by individuals committed to promoting Galileo as the uncontested father of modern science and founder of the experimental method. At the turn of the twentieth century, Antonio Favaro (1847–1922) arranged and printed them in the National Edition of Galileo's *Opere*, a collection that has been mined by generations of Galileo scholars.[6] Galileo's patrimony, filled with Galileo's own rhetoric, collected by Viviani to promote a specific vision of his teacher and then organized in the nineteenth and twentieth centuries, thus was deliberately designed to promote an image of Galileo as a novel scientist, one whose innovative ideas and methods set him apart from his peers.

The tendency of scholars to return to this same set of sources has reinforced the image of Galileo promoted by those who constructed the archive. The Galilean collection today numbers more than 50,000 papers distributed in 347 volumes, and it is thus not surprising that it is the principal collection to which scholars of Galileo have turned. This reliance on the same material has shaped the historiography of the *Two New Sciences* in ways that have instituted and perpetuated its image as a text of the "New Science."

Such a pattern can be seen clearly by tracing the intellectual heritage of work on the book's reception. Favaro set the tone for this scholarship by describing the intellectual community around Galileo as one divided between supporters and critics. His many publications include two collections of biographies and descriptions of archival materials associated with Galileo's friends, correspondents, and adversaries. The first collection, entitled *Friends and Correspondents of Galileo*, deals with forty-one individuals supportive of Galileo's work, including Viviani and Mersenne. The second collection, *Opponents of Galileo*, describes individuals, including Bérigard, who objected to Galileo's findings in print.[7]

Our current picture of the reception of the *Two New Sciences* owes even more to two later scholars, Alexandre Koyré (1892–1964) and Stillman Drake (1910–1993). Best known for his sweeping narratives describing large-scale shifts in seventeenth-century science, Koyré authored two smaller studies in the 1950s tracing more concretely the fate of Galileo's science of motion. In his 1953 "Experiment in Measurement" Koyré outlined the efforts of Galileo and his readers, namely, Mersenne, Riccioli, and Huygens, to determine a quantitative rule of fall through experiment.[8] Koyré returned to the reception of Galileo's science of motion in a 1955 article focusing on the problem of free fall. In this piece Koyré discussed the efforts of seventeenth-century scholars to determine the trajectory of a falling body on a moving Earth.[9] More than thirty years later, Drake revisited many of these readers in the last chapter of his 1989 *History of Free Fall*, in which he studied the reception of Galileo's findings on falling bodies from 1632 to 1649. Drake explored the reactions of readers including Baliani, Mersenne, Gassendi, Le Cazré, Fabri, and Descartes.[10] More recent scholarship has drawn explicitly on the work of Koyré and Drake.[11] On the whole, these studies have accepted the conceptual framework and underlying questions introduced by their predecessors, categorizing Galileo's readers as "supporters," "opponents," or somewhere in between and focusing on attempts to validate or challenge Galileo's key findings. The self-referential nature of this scholarship has been reinforced by the fact that these historians have returned repeatedly to the same set of actors, whose responses or references to them survive in the Galileo patrimony first preserved by Viviani.

I have deliberately moved beyond these contours, bringing in a greater variety of historical actors, who enrich the accepted narrative of the text's reception. In contrast to historians' interest in Day 3, the readers discussed in the preceding chapters responded to all four days of the *Two New Sciences*. They read the book with a variety of textual tools, some drawn from traditional natural philosophy and from mixed mathematics, others reflecting new approaches to the natural world. Only one reader, pseudo-Baliani, annotated with the express purpose of interrogating Galileo's assumptions and the relationship between his claims and natural phenomena. The evidence suggests that an eclectic mix of textual approaches were employed in conjunction by all sorts of readers, both those who were supportive of novel approaches and those who were not.

Viviani and the Galileo patrimony, however, are only partly responsible for the perception that the *Two New Sciences* inspired readers primarily toward verification and extension of its findings. The state of scholarship on the *Two New Sciences* is the product of larger forces in the field of early modern history of science. First, methodological divisions within the field have helped to isolate scholars of Galileo's mechanics from the larger community. The result of historians' revisions of the discipline's traditional narrative has been a bifurcation in the field. More mathematical works central to the old narrative, like the *Two New Sciences*, are increasingly studied by a small group of technically proficient historians whose research focuses on the development of new theoretical concepts. Even in recent years, the majority of historians have examined the *Two New Sciences* from the perspective of the text's genesis, seeking to explain how and why Galileo developed the theoretical insights underlying it.[12] The wider community, in contrast, embraces novel questions and methodologies and applies them, for the most part, to texts, individuals, and subjects overlooked by the discipline's grand narrative, often ignoring once-canonical texts like the *Two New Sciences*. In rare instances, the two approaches collide, setting off reverberations that hint at the incommensurable paradigms operating within the discipline itself. Studying the reception of the *Two New Sciences* was a deliberate attempt to force such a clash between a darling of the traditional narrative and the approaches and methods of revisionist historians of science. Applying the approaches of historians of the book and of reading to the text forces us to remove Galileo as the principal protagonist and to view the book through the eyes of his contemporaries. This is because historians of reading explicitly acknowledge the limits of new ideas by emphasizing the agency of readers, rather than the authoritative author, in interpreting and transmitting texts.

The second factor that has shaped the historiography is the set of conceptual approaches historians have employed to explain scientific change. In recent decades, historians have developed an expanding set of rich tools to understand how innovation was generated, whether through travel, trade and commerce, the alchemical laboratory, artisanal workshops, or the notebooks of scholars. We have comparatively fewer models of how innovation was received. The dominant model today continues to be based on Thomas Kuhn's notion of incommensurable paradigms in conflict. This model allows us to use innovation as an explanation for intellectual transformation because it assumes no middle ground between adherents of novelty and of tradition. It creates a dichotomy between groups of historical actors, artificially constraining them to embrace or eschew new or old approaches.

Our histories of early modern science reflect these assumptions. Following Kuhn, we have privileged conflicts as favored episodes for historical scholarship for their ability to throw into sharper relief the contrast between opposing paradigms.[13] There has been a growing recognition that many period scholars, especially university professors and Jesuit writers, straddled the line dividing innovator from traditionalist. For the most part, however, historians have described these individuals as atypical or as intellectually weak. In the former sense, some have been portrayed as remarkable exceptions for their abilities to converse with competing camps.[14] The vast majority have been understood as important contributors to early modern science, to be sure, but as largely unfortunate souls, constrained by institutional regulations, the comprehensiveness of Aristotle's writings, the inertia of the pedagogical system, and religious vows and beliefs, so that their true loyalties—to innovation or tradition—remain obscure.[15]

The problem with this model is that it ignores the reality of the thinking reader, one who actively wanted to carry out good natural philosophy, to understand what he or she was reading, and to make sense of the swirl of new speculation being produced. Given the well-known difficulties faced by early modern readers and modern historians alike in interpreting Galileo's writings and the relationship between the practices he professed to embrace and those he actually employed, it makes much more sense to assume that most readers did come to his writings with no clear idea of what they should do with them. The most obvious response would have been to pick and choose, drawing on established techniques and also elements of Galileo's advice.

I have shown that Galileo's readers adopted just such an eclectic approach, suggesting that eclecticism, rather than binary division, was a convincing and widely used mechanism for responding to novel claims in the period. The university professors of chapters 5 and 6 claimed that they were creating new and improved philosophies—or just explicating Aristotle—but they included Galileo's work and that of other innovators alongside the opinions of Aristotle and his ancient, medieval, and early modern commentators. Their methods of incorporating Galileo into their expositions of Aristotle correspond with Charles Schmitt's vision of an eclectic Aristotelianism, whose openness to disparate and novel philosophical traditions made it more comprehensive.

Such a notion of an eclectic Aristotelianism, however, applies equally well to these readers' approach to Galileo. What they demonstrate is not just the eclecticism of early modern Aristotelianism but the intrinsic eclecticism of early modern philosophy more broadly. The openness of the dialogue genre and Galileo's use of it to communicate his own thoughts on both philosophical method and content may have encouraged such an approach.[16] As they read and wrote, these professors grappled with modes of producing knowledge that may seem contradictory to modern sensibilities: the bookish practices of traditional natural philosophy but also mathematical and experimental methods. Even the virtuosi readers of chapters 1–4, ostensibly supportive of the New Science, similarly read Galileo eclectically. Their annotations reveal that they drew on multiple intellectual traditions, including Aristotelian natural philosophy, practical mathematics, and new currents in experimental philosophy and physicomathematics, as they read the *Two New Sciences* and employed it for a diverse set of scholarly projects.

Recognizing the intellectual validity of the eclectic stance enriches our current models of the reception of novelty. It legitimates the role of the historical actors who retained ties to both novel and traditional claims. Explaining intellectual transformation according to this model also opens the door to new questions and approaches: When did readers recognize a given claim as necessitating new scholarly methods? How did old methods of reading facilitate or obscure new knowledge claims? What were the points of similarity between old and new scholarship that facilitated appropriation of the new? If we discard the assumption that old and new stood on opposite sides of an incommensurable divide, explaining change becomes an exercise in exploring the landscape of the middle ground between different approaches, not merely finding instances of innovation and contrasting them with established methods.

Galileo's readers indicated that a fruitful space for inquiry might lie in an expanded study of early modern textual practices. Historians have argued for a strong association between textual practices and the Baconian sciences.[17] Readers' reactions to Galileo's book suggest that these connections were not limited to descriptive fields but rather extended to Galileo's mathematical mechanics. In contrast to their rhetoric, Galileo and his readers continued to employ textual methods derived from traditional natural philosophy until the turn of the eighteenth century. These findings confirm the conclusions of scholars who have argued for the relevance of the humanist, textual tradition to seventeenth-century scholarship. In fact, they extend them to show that not only did the two cultures of humanism and science coexist and often collaborate but in many ways the dissemination and transformative power of the latter depended on the tools and methods of the former.[18]

Galileo and other period innovators called their contemporaries to arms with vivid images of an intellectual community divided between clear-sighted innovators and traditionalists committed to bookish scholarship, even when their own practices did not map neatly onto this vision. Their rhetoric was so persuasive that it has been difficult for historians to recover the voices of the many period actors for whom the intellectual community was not such a battleground. The situation is paradoxical. Galileo once proudly proclaimed the *Two New Sciences* as a replacement for the old-fashioned textual scholarship of Aristotle, yet now the text is largely considered passé by the larger historical community. Early modern readers made of the *Two New Sciences* many different things, constructing knowledge and meaning not by embracing Galileo's vision of a new science but by cobbling together a variety of approaches for diverse ends. They viewed the book as part of their wider scholarly projects, and it is time that historians did the same.

Introduction

1. Camerota, *Galileo Galilei e la cultura scientifica*, 537; Drake, *Galileo at Work*, 357–58; Heilbron, *Galileo*, 325–29; Wootton, *Galileo*, 215–17, 237. On Galileo's relationship with his daughter, see Sobel, *Galileo's Daughter*.

2. Camerota, *Galileo Galilei e la cultura scientifica*, 330; Drake, *Galileo at Work*, 441; Heilbron, *Galileo*, 330. On the relationship between Diodati and Galileo, see Garcia, *Élie Diodati et Galilée*.

3. Galileo's response to Antonio Rocco's criticism of his 1632 *Dialogue* has attracted the most attention. See Drake, *Galileo at Work*, 61–67, 359; Galilei, *Discorsi e dimostrazioni matematiche*, ed. Carugo and Geymonat (1958), esp. 623, 626–31, 637–38, 640–43, 659–61; and Galilei, *Le opere di Galileo Galilei*, ed. Favaro, Edizione Nazionale,7:569–750 (hereafter cited as Galilei, *EN*).

4. Camerota, *Galileo Galilei e la cultura scientifica*, 544–45; Drake, *Galileo at Work*, 374; Heilbron, *Galileo*, 347–48; van Helden, "Longitude and the Satellites of Jupiter," 92.

5. Camerota, *Galileo Galilei e la cultura scientifica*, 545–47; Drake, *Galileo at Work*, 373–74; Galilei, *EN*, 16:316, 326–27, 416.

6. For a brief overview of how turn-of-the-century historians of science portrayed Galileo in their narratives of the development of mechanics, see Gabbey, "Newton's Mathematical Principles of Natural Philosophy," 306–7.

7. Koyré's vision of the period is described and contextualized in Lindberg, "Conceptions of the Scientific Revolution," 15–18.

8. A. Rupert Hall, *Scientific Revolution*, 75–76.

9. Bertoloni Meli, "Mechanics," 659–64; Vilain, "Christiaan Huygens' Galilean Mechanics," 187–88.

10. For example, Heilbron terms it Galileo's "favorite" work. Heilbron, *Galileo*, 331.

11. Palmerino, "Galileo's and Gassendi's Solutions," 421.

12. For two such examples, see Biener, "Galileo's First New Science"; and Palmerino, "Una nuova scienza della materia."

13. For one example of this inclusiveness, see Park and Daston, *Early Modern Science*.

14. For an overview of these responses, see Bertoloni Meli, "Mechanics," 649–53.

15. For the text of the letter, see Mersenne, *Correspondance*, 8:94–133. For a partial English translation, see Descartes, *Philosophical Writings of Descartes*, 3:124–28.

16. Eisenstein, *Printing Press as an Agent of Change*, vol. 2, esp. 575–635; Johns, *Nature of the Book*, 40–48. For some key examples, see, in addition to those listed below, Frasca-Spada and Jardine, *Books and the Sciences in History*; Kusukawa, *Picturing the Book of Nature*; Needham, *Galileo Makes a Book*; Wilding, *Galileo's Idol*; and Yale, *Sociable Knowledge*.

17. Jardine and Grafton, "Studied for Action," 30–33.

18. See, e.g., Ariew, *Descartes and the Last Scholastics*; Des Chene, *Physiologia*; Dear, *Mersenne and the Learning of the Schools*; and Martin, *Renaissance Meteorology*.

19. Schmitt, "Renaissance Aristotelianisms"; Schmitt, "Eclectic Aristotelianism."

20. There is a large and growing body of literature devoted to the history of scientific practice. For an overview of recent work relating to early modern science, see Shapin, *Scientific Revolution*, 191–95.

21. On reading and note-taking as part of scientific practice, see Blair, "Rise of Note-Taking in Early Modern Europe."

22. Saenger, *Space between Words*, 6–13.

23. Grafton, "Humanist as Reader," 196–210.

24. Blair, "Rise of Note-Taking in Early Modern Europe," 309–12.

25. For an explicit discussion of the bookish nature of this textual enterprise, see Blair, *Theater of Nature*, 49–115; and Blair, "Humanist Methods in Natural Philosophy." On the fusion of Scholastic and humanist methods by the late sixteenth century, see Costello, *Scholastic Curriculum at Early Seventeenth-Century Cambridge*, 14–31; Cochrane, "Science and Humanism in the Italian Renaissance," 1055–57; and Grendler, *Universities of the Italian Renaissance*, 199–248.

26. On the notion of paper technologies, see Heesen, "Notebook."

27. Daston, "Taking Note(s)."

28. Blair, *Theater of Nature*, 3. For other examples (though this is not an exhaustive list), see Blair, *Too Much to Know*; Blair, "Note Taking as an Art of Transmission"; Blair, "Reading Strategies for Coping with Information Overload"; and Blair, "Rise of Note-Taking in Early Modern Europe."

29. On Julius Caesar, see Sherman, *Used Books*, 127–48. See also, e.g., Nelles, "Libros de Papel, Libri Bianchi, Libri Papyracei"; Sherman, *Used Books*, esp. 71–148; and Soll, "Amelot de La Houssaye (1634–1706) Annotates Tacitus."

30. The Newton Project, directed by Rob Iliffe and Scott Mandelbrote, http://newtonproject.sussex.ac.uk/.

31. For discussions of some of these surviving notes and books, see Mandelbrote, *Footprints of the Lion*, esp. 15–17, 69–70, 91–94, 117–19.

32. On Newton's use of books, see Harrison, *Library of Isaac Newton*, esp. 1–27.

33. On Boyle's textual practices in the realm of alchemy, see Principe, *Aspiring Adept*, 138–49. On his note-taking practices compared with contemporary practice, see Yeo, "Loose Notes and Capacious Memory"; and Yeo, *Notebooks, English Virtuosi, and Early Modern Science*, 133–73.

34. Newman and Principe, *Alchemy Tried in the Fire*, 91–94, 156–79.

35. Some key examples of studies of the reception of innovative, scientific texts in the period include Gingerich, *Annotated Census of Copernicus'* De Revolutionibus; Gin-

gerich, *Book Nobody Read*; Renn and Damerow, *Equilibrium Controversy*; Ariew, *Descartes and the Last Scholastics*; Bucciantini, Camerota, and Giudice, *Il telescopio di Galileo*; French, *William Harvey's Natural Philosophy*, 114–309; Guicciardini, *Reading the Principia*; and Jolley, "Reception of Descartes' Philosophy."

36. Darnton, "First Steps toward a History of Reading," 157.

37. Galluzzi, *Momento*; Giusti, "Master and His Pupils"; Wisan, "New Science of Motion"; Wisan, "On the Chronology of Galileo's Writings"; Wisan, "Galileo's Scientific Method"; Wisan, "Galileo's 'De systemate mundi' and the new Mechanics"; Damerow et al., *Exploring the Limits of Preclassical Mechanics*, 135–278.

38. Drake, *Galileo at Work*; Drake, *Galileo's Notes on Motion*; Settle, "Experiment in the History of Science"; Settle, "Galileo and Early Experimentation"; Settle, "Galileo's Use of Experiment as a Tool of Investigation"; Palmieri, *Reenacting Galileo's Experiments*; Renn, Damerow, and Rieger, "Hunting the White Elephant."

39. Büttner, Damerow, and Renn, "Traces of an Invisible Giant"; Valleriani, "View on Galileo's 'Ricordi Autografi.'"

40. Valleriani, *Galileo Engineer*, esp. 193–211.

41. Two examples are Butterfield, *Origins of Modern Science*, 17; and Koyré, *Galileo Studies*, 3.

42. For key examples, see Drake, *Galileo Studies*, esp. 95–99; Shea, *Galileo's Intellectual Revolution*, esp. viii; and Biagioli, *Galileo, Courtier*, esp. 159–244.

43. Jaki, introduction, xviii–xix; Duhem, *Medieval Cosmology*; Crombie, *Augustine to Galileo*; Schmitt, "Experience and Experiment"; Schmitt, "Faculty of Arts at Pisa at the Time of Galileo"; Crombie, "Sources of Galileo's Early Natural Philosophy"; Carugo and Crombie, "Jesuits and Galileo's Ideas of Science and of Nature"; Wallace, *Galileo and His Sources*; Helbing, *La filosofia di Francesco Buonamici*; Camerota and Helbing, "Galileo and Pisan Aristotelianism."

44. Galilei, *EN*, 6:232.

45. Ibid., 10:423.

46. Scholars have emphasized Galileo's study of natural philosophy as taught at the Collegio Romano and Pisa. For key examples of such work, see Camerota and Helbing, "Galileo and Pisan Aristotelianism"; Carugo and Crombie, "Jesuits and Galileo's Ideas of Science and of Nature"; Crombie and Carugo, "Sorting Out the Sources"; and Wallace, *Galileo and His Sources*.

47. Crystal Hall, *Galileo's Reading*, 14–43.

48. See Camerota, "La biblioteca di Galileo," and sources cited therein; and Crystal Hall, "Galileo's Library Reconsidered."

49. For various examples, see Biagioli, *Galileo's Instruments of Credit*, 7–8; Drake, *Galileo's Notes on Motion*; Galilei, *EN*, 7:569–750; and Renn, "Galileo's Manuscripts on Mechanics."

50. Raphael, "Printing Galileo's *Discorsi*"; Wilding, "Manuscripts in Motion."

51. Drake, *Galileo at Work*, 355–56. For the letters exchanged, see Galilei, *EN*, 15:274–75, 279–81, 283–84.

52. Drake, *Galileo at Work*, 376.

53. Westman, "Reception of Galileo's 'Dialogue,'" 348.

54. Galilei, *EN*, 16:30.

55. Ibid., 16:53, 61; Drake, *Galileo at Work*, 361.

56. Galilei, *Discorsi e dimostrazioni matematiche*, ed. Carugo and Geymonat (1958), xx.

57. Galilei, *EN*, 16:61.

58. Ibid., 16:140–41, translated in Drake, *Galileo at Work*, 362–63.

59. Drake, *Galileo at Work*, 361.

60. Galilei, *Discorsi e dimostrazioni matematiche*, ed. Carugo and Geymonat (1958), 622–23, 670–74.

61. Galilei, *EN*, 8:165. On the proportion of the manuscript in Galileo's hand versus that of the copyist, see ibid., 8:177.

62. De Ceglia, "Additio Illa Non Videtur Edenda," 160. De Ceglia's analysis examines Jesuit internal censorship records to reveal the connections Biancani saw between his own work and that of Galileo.

63. Clucas, "Galileo, Bruno, and the Rhetoric of Dialogue," 406; Crystal Hall, *Galileo's Reading*, 130–40; Jean Dietz Moss, *Novelties in the Heavens*, 257–300.

64. Nicholas Jardine, "Demonstration, Dialectic, and Rhetoric in Galileo's *Dialogue*."

65. Clucas, "Galileo, Bruno, and the Rhetoric of Dialogue," 411–12; Finocchiaro, *Galileo and the Art of Reasoning*, 27–28.

66. Raphael, "Galileo's *Two New Sciences* as a Model of Reading Practice."

67. Galilei, *EN*, 16:200–201, 203, 214, 218–28, 229–33. On De Ville's reaction, see Valleriani, *Galileo Engineer*, 124–25.

68. Galilei, *EN*, 16:209, 214.

69. For their reactions, see ibid., 16:209, 255, 268, 296, 510. Valleriani is one of the few scholars who identifies Tensini's birth and death dates (1580–1630). Valleriani, *Galileo Engineer*, 124. Micanzio, however, reports Tensini's reading of the *Two New Sciences* on 10 February 1635. Galilei, *EN*, 16:209. Marc Antonio Celeste also appears to have been a recipient of the manuscript. See ibid., 16:229.

70. Westman, "Reception of Galileo's 'Dialogue,'" 333–34. Mario Biagioli offers a similar interpretation based on Galileo's rhetoric. Biagioli, *Galileo, Courtier*, 216–18.

71. For an alternative interpretation of the audience of the *Two New Sciences*, see Raphael, "Making Sense of Day 1 of the *Two New Sciences*"; and Raphael, "Galileo's *Two New Sciences* as a Model of Reading Practice."

72. Various spellings of the name Elsevier were in use in the seventeenth century. Here I follow the lead of David Davies and spell it "Elsevier" even when variants are employed in the surviving correspondence. See Davies, *World of the Elseviers*, vi. On Galileo's correspondence with the Elseviers, see Favaro, *Amici e corrispondenti di Galileo*, 3:1377–1410; and Drake, *Galileo at Work*, 367–70, 373–74.

73. Mersenne's copy (BSG, 4V 585 INV 1338 RES), examined in chapter 3, and a copy, held at the Wellcome Library (EPB/2648/B), conform to this pattern. A cataloger's note suggests that this was an early issue of the book. The London-based book dealer W. P. Watson described a similar copy from the science collection of the head of the Royal Swedish Academy of Sciences as a surviving exemplar of the "rare and virtually unrecorded first issue" of the first edition of the *Two New Sciences*, though his description erroneously describes the collation following the standard printing (not the first issue). Watson also reports that another such copy was reported in Charles

Traylen's catalog 66, of 1966. Watson, *Science, Medicine, Natural History, Catalogue 19*, item 38.

74. There was much confusion regarding the contents of this book. A "Fifth Day" was published separately by Galileo's student Vincenzo Viviani in Florence in 1674 as *Principio della Quinta Giornata del Galileo* and included in the 1718 edition of Galileo's *Opere*. This 1718 edition also contained the "Sixth Day," on the force of percussion. For a list of editions of the *Two New Sciences*, see Galilei, *Two New Sciences*, 309–11. On the publication history of these later days, see Galilei, *Discorsi e dimostrazioni matematiche*, ed. Carugo and Geymonat (1958), 850–51.

75. Costabel and Lerner, introduction, 15. The 1638 edition of the *Two New Sciences* does not appear in the Easter or Michaelmas catalogs of the Frankfurt book fair from 1635 to 1639. Many thanks to Nick Wilding for sharing with me copies of these catalogs.

76. On this price and the salary of the Venetian official, see Shea, *Designing Experiments and Games of Chance*, 24.

77. Nuovo, *Book Trade in the Italian Renaissance*, 111.

78. Drake, *Galileo at Work*, 398.

79. Mersenne, *Correspondance*, 7:314; 8:64, 94–115, 173–77, 606–14. On Mersenne's access to the text before it was printed, see Costabel and Lerner, introduction, 20–28.

80. These distribution figures were obtained by my own partial census, carried out via e-mail and personal visits to libraries from 2012 to 2013. Copies with seventeenth-century inscriptions and/or bookplates include Merton College, Oxford, A9(2); Christ Church, Oxford, Allestree c.5.21; and Bodleian, Savile Bb.13.

81. BNCF B. Rari 169; BNCF Gal. 79.

82. Buccolini, "Opere di Galileo Galilei." Digby's copy is Bibliothèque interuniversitaire de la Sorbonne, RR 8=89.

83. On Wren's copy, see chapter 4. Duke August's copy is Herzog-August Wolfenbüttel Bibliothek, A: 33.9 Quod.

84. Galilei, *Opere* (1655–56), title page.

85. Galilei, *EN*, 8:23, 214. The efforts of Galileo's students and others to publish the 1655–56 edition are detailed in their surviving letters. For citations and a brief discussion of the role of the Pisan professor Carlo Rinaldini, see Baldini, "Tra due paradigmi?," 189–90.

86. On Mersenne, see Lenoble, *Mersenne, ou, La naissance du mécanisme*; and Dear, *Mersenne and the Learning of the Schools*. On the *Nouvelles pensées* specifically, see Raphael, "Galileo's *Discorsi* and Mersenne's *Nouvelles Pensées*"; and Shea, "Marin Mersenne."

87. Mersenne, *Les nouvelles pensées de Galilée*, 1:53.

88. Wilding, "Return of Thomas Salusbury's *Life of Galileo* (1664)," 242–43. Each of the two volumes was made up of two tomes.

89. Salusbury, *Mathematical Collections and Translations*, 2, pt. 1, title page.

90. Galilei, *Systema Cosmicum*, title page.

91. Galilei, *EN*, 8:27–28. For a list of these editions, see Galilei, *Two New Sciences*, 309–10.

92. Bertoloni Meli, *Thinking with Objects*, 117–20.

93. *Compleat gunner.* The title page explains that to this work "is added The doctrine of projects applyed to gunnery by those late famous authors Galilaeus and Torricellio now rendred into English."

94. Blondel, for example, mentioned his intention to publish a work entitled "Galilaeus Promotus de Resistentia Solidorum," which was never printed. Marchetti claimed that his 1669 *De Resistentia Solidorum* had originally been titled *Galilaeus Ampliatus.* In 1718 Grandi saw into print Viviani's *Trattato delle resistenze,* which was an attempt to discuss and illustrate Galileo's science of the resistance of materials. For these, see Benvenuto, *Introduction to the History of Structural Mechanics,* 1:233–52.

95. The subject bibliography devoted to philosophy by Martin Lipen (1630–1692), professor in the gymnasium of Halle and of Stettin and later joint rector of the Academy of Lübeck, for example, mentioned neither the *Two New Sciences* nor the *Dialogue,* though it did list the Latin translation of Galileo's *Letter to Christina.* It also mentioned a treatise identified as *Disputations on Lux and Lumen,* ostensibly printed in 1612 in Venice and composed by Galileo. On Lipen's project, see Balsamo, *Bibliography: History of a Tradition,* 80. The translation of Galileo's letter was listed under the heading "Testimonia Scripturae." Lipen, *Bibliotheca Realis Philosophica,* 2:1471. "Gal. de Galilaeis Disp. De Luce et Lumine. Venet. 4. 1612" was listed twice under the headings "Lumen. conf. Lux" and "Lux. conf. Lumen." Ibid., 1:857, 862.

96. *Catalogus Librorum Diversis Italiae Locis Emptorum Anno Dom. 1647,* 36.

97. *Bibliotheca Hookiana,* 8. The section containing Italian books is found in the appendix, 12–17. The *Bibliotheca Hookiana* is also reprinted in Rostenberg, *Library of Robert Hooke,* 141–221; and Feisenberger and Munby, *Sale Catalogues of Libraries of Eminent Persons,* 11:37–116.

98. Mersenne, *Correspondance,* 12:220. That Mersenne refers here to the *Two New Sciences* is clear from the content of the letter, which addresses specific pages from the work.

99. Hyde, *Catalogus Impressorum Librorum Bibliothecae Bodleianae in Academia Oxoniensi,* 274.

100. Holstenius, *Index Bibliothecae qua Franciscus Barberinus, S.R.E.,* 1:445.

101. For some examples, see Dinis, "Was Riccioli a Secret Copernican?"; Siebert, "Kircher and His Critics"; Stolzenberg, "Oedipus Censored"; and Gorman, "Jesuit Explorations of the Torricellian Space."

102. See, e.g., Baldini, *Legem Impone Subactis,* 75–119, 251–81; Dinis, "Was Riccioli a Secret Copernican?"; Hellyer, "Because the Authority of My Superiors Commands"; and Hellyer, *Catholic Physics,* 35–52.

103. In specific instances, as described in chapter 4, individuals may have aligned themselves in such camps. For another example, see Roux, "Empire Divided."

104. See, e.g., Galileo's *Letter to Christina,* Galilei, *EN,* 5:309.

105. "Quum autem variae, atque pugnantes inter se Philosophorum sententiae, hanc scientiam perdifficilem, atque adeo perplexam effecerunt." BML, Giannetti, "Physica," 1.

106. For these usages, see, e.g., Dear, *Mersenne and the Learning of the Schools;* Feingold, *New Science and Jesuit Science;* and Hellyer, *Catholic Physics.*

107. Bérigard, *Circulus Pisanus* (1643), 1:59, 60; 3:32. For a Jesuit example, see Mauro, *Quaestionum Philosophicarum . . . Libri Tres,* 3:400. These Jesuit philosophers also tended to identify Galileo as a follower of Copernicus in those sections of their

commentaries devoted to Aristotle's *On the Heavens*. For example, Cattaneo, *Cursus Philosophicus*, 2:753–54.

    108. Aristoteles restitutus Philosophia de Veteris Nova." BML, Mancini, "De Physica," 51r.

Chapter 1: An Anonymous Annotator, Baliani, and the "Ideal" Reader

    1. Sixty-nine of 107 pages are annotated in Day 1; 27 of 41 in Day 2; 43 of 86 in Day 3; and 29 of 52 in Day 4.

    2. Procissi claims that the catalog drawn up by Antonio Favaro and Alarico Carli and titled "Indice analtico dei Manoscritti Galileiani" names Torricelli as a potential annotator. Procissi, *Collezione Galileiana*, 1:160. In his own discussion of the annotated volume in the introductory comments to the *Two New Sciences* in his edited *Opere* of Galileo's works, Favaro makes no such suggestion. See Galilei, *EN*, 8:24n7.

    3. Drake, *Galileo at Work*, 231, 250, 439.

    4. Galilei, *EN*, 8:197–98.

    5. "Nota quod haec est definitio motus aequabilis, quae constat quibus de facto dari. Datur ibidem ab auctore definitio motus uniformiter accelerati, at non constat de facto dari, unde qui eum vult asserere, debet et . . . [illegible] probare aut experientia aut ratione demonstrativa non autem probabili, si vult ut id indubitanter credamus." BNCF Gal. 80, 157. Transcriptions of marginal annotations follow the grammar and spelling of the original; grammatical errors have been indicated only when translation to English in the text necessitated commentary. I have silently expanded standard abbreviations in manuscript treatises.

    6. Dear, *Discipline and Experience*, 23.

    7. "Questo è tutto il punto, cioè se tal definizione si adatti a quel moto accelerato con cui i gravi naturalmente descendono il che l'Autor presuppone." BNCF Gal. 80, 158.

    8. "È probabile ma non certo potendo esser che nelle sesta fossero tre, o cinque, e cosi nelle altre, onde non so' se tal proposizione fosse ammessa per principio è una scienza, e di dimostrare." Ibid., 159.

    9. "È però vero che questo trattato dipende da quei principii e supposti, o petitioni come nelle note in margine si può vedere, onde non so io come con verità si possa dire che ogni cosa si diduca da un solo semplicissimo principio. Ma quel, che è più è che quello tal principio non è altro che la definizione del moto accelerato uniformemente, nel [proposto] poi si suppone che il moto naturale sia tale, cioè uniformamente accelerato, però questo non si pruova, come il pruovo io del moto à piombo alla terza proposizione e dell'inclinato alla seconda, onde questo autore fa come colui chi supporta la quadratura del circulo, ne . . . [illegible] le passioni, e ne deduce molte bellisssime conclusioni, e percio sono ambidue degni di molte lode ma non pertanto è vero che non si pruova il punto principale cioè che i corpi naturali di moto naturale habbiano e sien fatti con le proporzioni che qui si asserisce, ma solo che le havrebbono se in fatto si movessero con moto che realmente si andasse accelerando uniformamente, il che ho' provato io ne luoghi prima accennati." Ibid., 235. The reader references his marginal annotations to Propositions 2 and 3 to this section of Day 3. In these annotations, the reader adds to Galileo's statements of the propositions by indicating that the propositions assume Galileo's definition of accelerated motion.

10. "La dimostrazione procede supposto che i due moti si mantengono dello stesso tenore, come se fossero ogn'un di loro da per se, e non si diano impedimento alcuno l'uno all'altro, dal quale impedimento risulti un altra ancor specie di moto, che à la sezione prima f. 236. Io pero dubito assai se tal supposto sia vero, anzi credo tutto'l contrario perciò che se fosse vero, il proietto crescerebbe tuttavia di velocità, e farebbe maggior percossa, ove io con la balestra ho' osservato che . . . [illegible] fa percossa maggiore, ne ciò può procedere solo dell'aria, perciò che nel moto perpendicolare in distanze di 50 piedi l'aria non causa differenza sensibile, e nel proietto si vede differenza notabile in distanze molti minore, onde . . . [illegible] pure da credere che il moto naturale dia grande impedimento al traversale." Ibid., 243.

11. "Più mi da noia il parermi potersi dubitare che nel proietto quei due moti si dessero tal impedimento l'uno all'altro, che si alterassero, non servendo quella tal proporzione che havrebbbero se fossero l'un senza l'altro, onde ne risultasse impedimento al moto, et all'impeto da esso causato." Ibid., 246.

12. Blair, *Theater of Nature*, 49–81.

13. Gingerich and Westman, "Wittich Connection," 28, 36; Westman, *Copernican Question*, 259.

14. Raphael, "Galileo's *Two New Sciences* as a Model of Reading Practice."

15. Baliani, *De Motu Naturali Gravium Solidorum*; Baliani, *De Motu Naturali Gravium Solidorum et Liquidorum* (1646). In the citations that follow, I reference the modern Latin edition and Italian translation, Baliani, *De Motu Naturali Gravium Solidorum et Liquidorum* (1998). I thank an anonymous reviewer for this suggestion.

16. Baliani, *De Motu Naturali Gravium Solidorum et Liquidorum* (1998), 263–65, esp. 263.

17. Ibid., 266, 272.

18. Ibid., 70–71. On the influence of Fabri, see the following and the sources they cite: Galluzzi, "Gassendi and l'Affaire Galilée of the Laws of Motion," 265–69; Baliani, *De Motu Naturali Gravium Solidorum et Liquidorum* (1998), 41–46.

19. Baliani, *De Motu Naturali Gravium Solidorum et Liquidorum* (1998), 140–45.

20. "Supposito motu unif. accelerato qui si deretur haec passio se queretur, et hic nullibi probatur de facto dari, quod ego probavi in meo tractatu de Motu naturali." BNCF Gal. 80, 171.

21. Blair, *Too Much to Know*, 104–12. Such an identification is also problematic given the use of different hands in carrying out different genres of scholarly work (annotating versus letter-writing, etc.).

22. Bertoloni Meli, *Thinking with Objects*, 105–34.

23. Ibid., 107, 111, 113–16, 126–34.

24. Bertoloni Meli, "Mechanics," 650.

25. See, e.g., Galilei, *EN*, 14:342; 18:75–79; 18:102–3.

26. On Galileo's relationship with Sagredo, see Wilding, *Galileo's Idol*.

27. Galilei, *EN*, 16:300–302.

28. Ibid., 7:47–53.

29. Ibid., 7:190.

30. Ibid., 7:489.

31. For two examples, see Clavelin, *Natural Philosophy of Galileo*, esp. 221–23; and Westfall, *Force in Newton's Physics*, 18.

32. Galluzzi, "Gassendi and l'Affaire Galilée of the Laws of Motion," 239. See also Galluzzi, "Gassendi e l'affaire Galilée delle leggi del moto."

33. BNCF Gal. 80, 243. See above, n. 10.

34. "Non perche l'aqua pesi 400 volte più dell'aria ma solo 50, perciò che cadendo due gravi cioè uno di piombo l'altro uguale all'acqua, quanto in 50 piedi restando circa un piede onde se tanto si detrae di velocità come di peso come a f. 77 l'aria vien e detrar di peso all'acqua 1/50 solamente." BNCF Gal. 80, 81. Page 77 of the 1638 edition corresponds to *EN*, 8:120. The reader actually refers to a marginal annotation found on page 76 of his copy.

35. Park, "Observation in the Margins," 29–30.

36. Gingerich and Westman, "Wittich Connection," 32.

37. Hess and Mendelsohn, "Case and Series," 287–91; Pomata, "Observation Rising."

38. Newman and Principe, *Alchemy Tried in the Fire*, 174–95; Leong, "Herbals She Peruseth."

39. "a per 3 huius." BNCF Gal. 80, 199.

40. Galilei, *EN*, 8:215.

41. "Il momento della pietra A al momento in G hà proporzione composta di GN à NC e di FB a BO, sia come FB a BO, cosi NC à X, è il Momento della pietra, à G come GN ad X, perche si Potenza B alla C e' come FO à OB, e componendo il momento A alla potenza G è come FB à BO, [come] NC à X. La potenza C alla G è come GN à NC dunque il momento A à G come GN à X." BNCF Gal. 80, 113.

42. "Non sarebbe terminabile ma si potrebbe far in infinito sincategorice, come si dice nelle scuole ma non mai sarebban in infinite in atto." Ibid., 34.

43. On the terms *categorematic* and *syncategorematic* in medieval Scholastic logic, see Kretzmann, "Syncategoremata, Exponibilia, Sophismata." For the terms in relation to medieval discussions of the infinite, see Murdoch, "Infinity and Continuity," esp. 567–68. On early modern treatment in the Jesuit context, see Raphael, "Making Sense of Day 1 of the *Two New Sciences*," 484–85.

44. With respect to Galileo's comment about the continuum being composed of infinite indivisibles, the reader noted, "Però convien provarlo" and "Vedremo però come si pruovi." BNCF Gal. 80, 32. With respect to the statement about unity, he wrote, "Se fosse stato provato—d'anzi che l'infinito numero è in atto, e che di qui forse ne risulterebbe che e' l'unità, ma si niega che l'infinito si dia, o dar si possa." Ibid., 38. On the division of a line into indivisible parts, "Qui sta il punto che convien provarlo." Ibid., 36.

45. "Meglio il dire che son infinite sincategorice." Ibid., 36.

46. "Può esser che l'acqua sia condensabile e rarefabile come l'aria, se ben meno, e che attraga con . . . [illegible] nella tromba si vada rarefacendo più quanto si attrae più in alto, e che in altezza di braccia 18 sia rarefatta quanto e' atta a rarefarsi onde alzata piu si spazi." Ibid., 17.

47. "I fili difficilmente strisciando si separano per lo contrasto che fan le parti fra loro, il che è causa che le pietre o' altro mal si tirano per terra senza aiuto di ruote, o mulinelli, e questa è la vera causa che la funa e suoi filamenti non si rompano, e che si trattenga coinvolta intorno all'argano ancorche si alzino pesi gravi. Il detto contrasto dà tal impedimento perciò che le parte han ruvidezza. ciò è di . . . [illegible] qualità almeno piccole, le quali entrando quelle dell'uno in quelle dell'altro, non ne puon uscire

senza quale lo moto per lo traverso, anzi più moti per lo traverso i quali si' impediscono l'un l'altro onde le funi liscie come vetro, scorrerebbeno." Ibid., 9.

48. On this notion of "experience" and developing notions of modern experiment, see the following and the sources they cite: Baroncini, *Forme di esperienza e rivoluzione scientifica*, esp. 39–62; Dear, *Discipline and Experience*, 21–25; Daston, "Baconian Facts, Academic Civility, and the Prehistory of Objectivity"; Schmitt, "Experience and Experiment."

49. "Credo che sia per dissensione dell'humido non con l'aria, ma col secco." BNCF Gal. 80, 71.

50. "Sia tavola o pietra AB sopra di cui la gocciola CD si sostiene perciò che altrimenti converrebbe che le parti C e D scorressero verso A e B, e perciò sopra al secco, loro contrario. Overò [*sic*] perche [*sic*] la Tavola sia ruvida, sopra di cui l'acqua habbia dificulta' a scorrere e le sia men dificile il sostenersi tale, che superar della dificultà di scorrere della ruvidezza." Ibid. The reader's discussion in terms of a table (*tavola*) rather than a cabbage leaf (*foglia del cavalo*), which is how Galileo introduced the phenomenon, may suggest familiarity with Galileo's *Discourse on Floating Bodies*, in which he addressed the behavior of *tavolette* floating on or submerged in water. See, e.g., Galilei, *EN*, 4:98.

51. "Dirai l'acqua E dovrebbe scorrere sopra la superficie EC, overo ED, rispondo che l'acqua E come grave non tende in C ne in D, ma al centro del mondo anche la trattiene il convenir alle parti C e D muoversi verso A e B per la causa detta. Ne vale che vicino a E fosse una palla e'l CED fosse un solido, che essa palla scorrerebbe sopra la superficie EC, perciò che è per esser il centro della gravita d'essa palla, sul . . . [illegible] e l'acqua che scorre sopra va piano pendente è per scendere." BNCF Gal. 80, 71. Part of the reader's response at the bottom of the page was cut off when the page was trimmed by a later owner or reader.

52. On the Scholastic *quaestio*, see Brockliss, "Curricula," 565–69, 578–89; and Marenbon, *Later Medieval Philosophy*, 10–14, 27–34. On pedagogical dialogues, see Blair, *Theater of Nature*, 52–55. For an example of a popular pedagogical dialogue and discussion of its content and genesis, see Cunningham and Kusukawa, *Natural Philosophy Epitomised*, esp. ix–xiii.

53. For discussions of this practice, see, e.g., Blair, *Theater of Nature*, 198–99; and Sherman, *John Dee*, 70–73, 82–83.

54. Westman, "Reception of Galileo's '*Dialogue*.'"

55. Mersenne's reading and citation of the *Two New Sciences* is discussed further in chapter 3.

56. Gassendi, *Opera Omnia*, 3:478–563.

57. An example is, "Causam duplicem Galileus assignat." Ibid., 3:496.

58. Ibid., 3:496, 498, 517. Though Galileo also discussed pendulums in the *Two New Sciences*, the problem of the pendulum bobs coming to rest is peculiar to the *Dialogue*. For the passage, see Galilei, *EN*, 7:256–57. On its particularity to the *Dialogue*, see Palmieri, "Phenomenology of Galileo's Experiments," 496–97.

59. Carla Rita Palmerino has argued that in his *De Motu*, Gassendi relied on the *Two New Sciences* to describe the trajectory of a ball falling from the mast of a ship as parabolic, not circular. Palmerino claims that Gassendi would have obtained this result from the *Two New Sciences*, since the only analogous situation in the *Dialogue* (a stone falling from a tower) is described by Galileo as following a circular motion. Palmerino,

"Galileo's Theories of Free Fall and Projectile Motion," 138, 145–46. As Palmerino herself notes, however, even in the passage supposedly corrected by his reading of the *Two New Sciences* Gassendi was "not completely faithful to the *Discorsi*." Ibid., 146. Finally, by the time his *De Motu* was published, Gassendi could have read other discussions of Galileo's parabolic trajectory in sources including Mersenne's *Nouvelles pensées* and Cavalieri's *Specchio ustorio*. For the latter, see Cavalieri, *Lo specchio ustorio*, 165.

60. Gassendi, *Opera Omnia*, 6:448–52.

61. Ibid., 6:448.

62. Ibid., 6:448–49.

63. Ibid., 6:448.

64. Borgato, "Riccioli e la caduta dei gravi"; Galluzzi, "Galileo contro Copernico."

65. See, e.g., Riccioli, *Almagestum Novum*, 1:55; references are to part and page.

Chapter 2: Editing, Commenting, and Learning Math from Galileo

1. Cavalieri, *Lo specchio ustorio*, 158; Drake, *Galileo at Work*, 340.

2. The distribution of surviving period manuscripts testifies to its diffusion. See Galilei, *EN*, 2:149–50. For more on its translation into French, see chapter 3. On the treatise in England, see Feingold, "Galileo in England," 416. Sophie Roux also kindly communicated to me her research on these surviving manuscripts.

3. For an overview of Torricelli's contributions, see Bertoloni Meli, *Thinking with Objects*, 119–20.

4. Ibid., 118–19.

5. Galilei, *EN*, 8:26–28.

6. Ibid., 8:27–28.

7. Findlen, "Living in the Shadow of Galileo."

8. Ibid., 244–51; Wilding, "Return of Thomas Salusbury's *Life of Galileo* (1664)," 252–54; Galluzzi, "Sepulchers of Galileo."

9. The copy is BNCF Gal. 79. Viviani's annotations were published in Galilei, *Discorsi e dimostrazioni matematiche*, ed. Giusti (1990). In the section that follows, citations refer to page numbers in the original annotated copy; when it is necessary to refer to the loose sheets bound in the volume, I rely on the folio numbers written in pencil in the copy. Antonio Favaro also discussed BNCF Gal. 79 briefly in his notes on the *Two New Sciences*. Galilei, *EN*, 8:25.

10. Viviani's identity as annotator has been confirmed and assumed by multiple scholars of Galileo. See, e.g., Galilei, *Discorsi e dimostrazioni matematiche*, ed. Giusti (1990), lxi–lxiii; Galilei, *EN*, 8:25.

11. "Per la faccia 114 dell'edizione di Leida 1638 e per la faccia 86 dell'edizione di Bologna 1656." BNCF Gal. 79, fol. 72r, slip between pp. 114 and 115.

12. James Hankins, *Plato in the Italian Renaissance*, 1:18–26.

13. Jardine and Grafton, "Studied for Action."

14. "Nei Discorsi e dimostrazioni matematiche di Galileo Galilei dell'impressione di Leida del 1638." BNCF Gal. 79, fol. 82, slip between pp. 128 and 129.

15. Early on in Day 1, for example, Viviani suggested that additional measurements be taken to improve Galileo's proposed method of measuring the force of the void: "di quel tal toccamento del zaffo con la superficie del cilindro: la qual resistenza di toccargli puossi misurar prima separata, con osservare quanto peso si richerchi a muover lo zaffo per lo

cannone quando il foro è aperto"; "detraendo quel peso (misura della resistenza dell'interno toccamento) dal rimanente peso, averemo"; "avanti la separazione del zaffo dall'acqua per la propria gravità, e per quella che se gli va aggiungendo del recipiente." BNCF Gal. 79, 16. Each of these annotations appears on the page twice, once in pencil and once in ink.

16. See, e.g., ibid., 194.

17. These notes are found in the third section of BNCF Gal. 74, 16r–39v, and labeled in the manuscript with the heading "Alcuni aggiunti di mano del Viviani. . . . "

18. "C. 254. il pensiero di Platone . . . C. 56. la dimostrazione del Torricelli de Cilindi . . . C. 81 nel modo di pesar l'aria si ha non solo il peso di essa nel vacuo, ma dell'acqua ancora nel medesimo: cosa non avvertita dal Galileo." BNCF Gal. 74, 33r–v. In his copy, Viviani offered an alternative demonstration on page 56 and inserted a loose sheet with a lemma and theorem and their accompanying proof following page 56. The provenance of these insertions (whether they are by Torricelli or Viviani) is not indicated in the annotated copy.

19. See the overview and specific studies provided in Blair, "Note Taking as an Art of Transmission," 92.

20. Nelles, "Libros de Papel, Libri Bianchi, Libri Papyracei," 84–85, 96–97.

21. Renn and Damerow, *Equilibrium Controversy*, 207, 214–19.

22. Oosterhoff, "Book, a Pen, and the Sphere."

23. Gingerich and Westman, "Wittich Connection," esp. 27–41.

24. Viviani's additions appear in BNCF Gal. 79, fols. 49v, 188r–189v. For transcriptions, see Galilei, *EN*, 8:21–22, 344, 437–38, 441. On Cosimo di Vincenzio and Viviani's interactions with him, see ibid., 8:27, 439n1, 452, 561; 19:15, 438n3. On Ferri, see ibid., 8:567; 19:439.

25. Gingerich and Westman, "Wittich Connection," 27.

26. Calis, "Personal Philology."

27. These notes are found in the same manuscript collection as those that contain Viviani's reading notes on the *Two New Sciences*, BNCF Gal. 74, 26v.

28. "Fusi." BNCF Gal. 79, 20.

29. "Atomi." Ibid., 49.

30. "Tritono o semidiapente," "diapason, dupla, ottava," "Diapente. Quinto. Sesquialtera," and "Diatesseron." Ibid., 104. Viviani's annotations are found in the passage corresponding to Galilei, *EN*, 8:147. The definitions of these chords are provided by Galileo in ibid., 8:143, 147.

31. The copy is Biblioteca Centrale di Ingegneria del Politecnico di Milano, Fondo Brioschi A III 00149.

32. "In Archimede nella proposizione 32 del primo libro *De sphaera et cylindro*, e più universalmente nella 29 *De conoidibus et sphaeroidibus*, ed altrimenti ancora." BNCF Gal. 79, 30.

33. "Sed hoc per se satis liquet. Nam extremae trium continue proportionalium simul sumptae maiores sunt duplae mediae geometrice proportionalis ex 25ª quinti Elementorum." Ibid., 216.

34. Ibid., 249. The two passages in question are Galilei, *EN*, 8:136–37 and 278–79.

35. "Hoc enim ostendit Galileus ipse in suo Mechanicae tractatu." BNCF Gal. 79, 181. The passage annotated by Viviani is Galilei, *EN*, 8:221.

36. Blair, *Too Much to Know*, 71.

37. Sherman, *John Dee*, 68–69.

38. Kusukawa, *Picturing the Book of Nature*, 255.

39. Renn and Damerow, *Equilibrium Controversy*, 182, 238–40.

40. Gingerich and Westman, "Wittich Connection," 28.

41. BNCF Gal. 79, 266.

42. For an example of an additional corollary, see ibid.

43. Gingerich and Westman, "Wittich Connection," 31–33, 48–49, 128–40.

44. "Per conversum secundae huius"; "per conversum propositionis primae huius." BNCF Gal. 79, 154.

45. Renn and Damerow, *Equilibrium Controversy*, 186–87.

46. For one example, see Bodleian, Savile X.9(2), 11r.

47. Marr, *Between Raphael and Galileo*, 91–93.

48. Groote, Kölbl, and Weiss, "Evidence for Glarean's Music Lectures from His Students' Books," 293–95, 299–302.

49. Gingerich and Westman, "Wittich Connection," esp. 32. Gingerich's census describes and offers illustrations of such diagrams. See, e.g., Gingerich, *Annotated Census of Copernicus' De Revolutionibus*, 12–13, 148, 151, 180, 184–85, 289.

50. Described in Westman, "Reception of Galileo's 'Dialogue,'" 347.

51. Gingerich and Westman, "Wittich Connection," 36.

52. For a summary of this scholarship, see Raphael, "Making Sense of Day 1 of the *Two New Sciences*," 489.

53. Gingerich and Westman, "Wittich Connection," 28.

54. Baldini, "Tra due paradigmi?," 189n3; Bertoloni Meli, *Thinking with Objects*, 118–19.

55. Grafton, *Culture of Correction in Renaissance Europe*, 23–32.

56. BNCF Gal. 79, title page (fol. 3r).

57. "Sentimento Eroico, e di Filosofo più che Cristiano, perché Cattolico e Santissimo, e non di'Ippocrita." Ibid., 31.

58. "Promessa d'altre speculazioni del Galileo." Ibid., 165.

59. "Quale possa esser la causa della coërenza delle parti de' corpi solidi." Ibid., table of contents (fol. 6v).

60. Ibid., 236.

61. Ibid., 235.

62. For example, "Le replicate sperienze colle quali ciascuno può soddisfarsi, mi hanno dimostrato che." Ibid., 96.

63. Dear, *Discipline and Experience*, esp. 124–29.

64. See BNCF Gal. 79, 127ff.

65. See, e.g., ibid., 297–98, 300–309.

66. On this question of fixity, see Johns, *Nature of the Book*, 31.

67. Yale, "Marginalia, Commonplaces, and Correspondence," 196.

68. Viviani's notes on the resistance of bodies to breaking are contained in BNCF Gal. 218, "Discepoli di Galileo, Tomo CVIII: Viviani, Vincenzio. Parte 3ª Meccanica dei Solidi. Volume 4. Proposizioni, Corrollari e Statica," and 219, "Discepoli di Galileo. Tomo CIX: Viviani, Vincenzio. Parte 3ª Meccanica dei Solidi. Volume 5. De solidorum mensura ac centro gravitatis." Examination of these manuscripts does not seem to reveal any treatise that specifically corresponds to the material found in the *Trattato*.

69. Grandi and Viviani, *Trattato delle resistenze*, 9.

70. Ibid.

71. "Della superiore." BNCF Gal. 79, 12.

72. August's behavior is described in Blair, "Rise of Note-Taking in Early Modern Europe," 313. Blair's account draws on Hess, "Fundamenta fürstlicher Tugend," 131–74.

73. See Paul Grendler's description of the techniques advocated by the sixteenth-century humanist Latin teacher Antonio della Paglia in Grendler, *Schooling in Renaissance Italy*, 245. For another instance of teaching ancient literature at the University of Paris, see Grafton, "Teacher, Text and Classroom," 41–42.

74. Grandi and Viviani, *Trattato delle resistenze*, 9.

75. Galilei, *EN*, 8:157.

76. See, e.g., BNCF Gal. 79, 116–19, 121–22, 124–26, 128, 131–32, 136, 140, 147–49.

77. Grandi and Viviani, *Trattato delle resistenze*, 4.

78. Galilei, *EN*, 8:151. On the relationship between these two sections of Galileo's text, see Biener, "Galileo's First New Science."

79. Grandi and Viviani, *Trattato delle resistenze*, 5.

80. Galilei, *EN*, 8:62–63.

81. On this practice in mathematical genres, see Gingerich and Westman, "Wittich Connection," 34. For a more general discussion, see Blair, *Theater of Nature*, 198–99; and Sherman, *John Dee*, 70–73, 82–83.

82. BNCF Gal. 79, 249.

83. See, e.g., ibid., 16, 112–15.

84. Grandi and Viviani, *Trattato delle resistenze*, 14.

85. Ibid.

86. On the conceptual origins of this Fifth Day, its intellectual interventions in the interpretation of Euclid, its publication, and its reception, see Giusti, *Euclides reformatus*, esp. 57–82.

87. On the publication of the "Universal science of proportions," which analyzes Viviani's rendering and misrenderings of Galileo's Fifth Day and work by Torricelli in the treatise, see ibid., 135–42.

88. Viviani, *Qvinto libro degli Elementi d'Evclide*, letter to the reader, recto.

89. Ibid., letter to the reader, verso to recto of following unnumbered page.

90. Ibid., 19.

91. Ibid., 68.

92. Ibid., 3.

93. Ibid., 4, 30.

94. Ibid., 2, 15, 42.

95. Ibid., 50.

96. On the Galileo school, see Pepe, *Galileo e la scuola galileiana*. On Viviani's membership in the school, see, e.g., Giusti, "Galileo all'origine delle ricerche della scuola galileiana," 8.

Chapter 3: Modifying Authoritative Reading to New Purposes

1. This introduction to Mersenne draws specifically on Garber, "On the Frontlines of the Scientific Revolution," 135–37, 142–45; and Bertoloni Meli, *Thinking with Objects*,

105–17, 130. The two most important studies of Mersenne are Dear, *Mersenne and the Learning of the Schools*; and Lenoble, *Mersenne, ou, La naissance du mécanisme*. On Mersenne's relationship to Galileo, see John Lewis, "Mersenne as Translator and Interpreter"; Palmerino, "Infinite Degrees of Speed"; Raphael, "Galileo's *Discorsi* and Mersenne's *Nouvelles Pensées*"; and Shea, "Marin Mersenne."

2. Mersenne, *Cogitata Physico-Mathematica*, "Phaenomena Mechanica," 63, 65.

3. Ibid., "Phaenomena Mechanica," 63–72, and "Ballistica," 93–94, 103–6.

4. Mersenne, *Les nouvelles pensées de Galilée*, 66. Here and below, citations of Mersenne's text correspond to the page numbers of the original edition, as they are reproduced in Mersenne, *Les nouvelles pensées de Galilée*, ed. Costabel and Lerner, vol. 1.

5. Bertoloni Meli, *Thinking with Objects*, 111. For Mersenne's discussions of these issues, see Mersenne, "Traité des mouvemens et de la cheute des corps pesans," 46–47; Mersenne, *Harmonie universelle*, 1:85–88; and Mersenne, *Novarum Observationum Tomus III Physico-Mathematicarum*, 192.

6. Garber, "On the Frontlines of the Scientific Revolution," 142–43.

7. Mersenne, *Correspondance*, vol. 5, app. 2.

8. Galilei, *EN*, 19:360–88; Finocchiaro, *Retrying Galileo*, 26–42.

9. Camerota, *Galileo Galilei e la cultura scientifica*, 330; Drake, *Galileo at Work*, 441; Heilbron, *Galileo*, 330. On Diodati's efforts specifically, see Gardair, "Elia Diodati e la diffusione europa del *Dialogo*."

10. Finocchiaro, *Retrying Galileo*, 43–64.

11. See Waquet, *Le modèle français et l'Italie savante*, 253–96, 389–439. On specific responses to Galileo, see Garcia, *Élie Diodati et Galilée*, 231–363; and Miller, *Peiresc's Europe*, 42, 86, 90, 109, 144–45. On Galileo's Italocentric outlook, in contrast, see Garcia, *Élie Diodati et Galilée*, 257–65. On the reactions of French readers to the *Dialogue*, see Beaulieu, "Les réactions des savants français au début du XVIIe siècle."

12. Bertoloni Meli, *Thinking with Objects*, 108–17, 125, 130.

13. Galluzzi, "Gassendi and l'Affaire Galilée of the Laws of Motion," 240.

14. The identification of this copy and a description of its annotations are found in Buccolini, "Opere di Galileo Galilei."

15. BIF, 4° M 541, 8, 169, 174.

16. BSG, 4V 585 INV 1338 RES.

17. Marie-Hélène de La Mure, e-mail message to author, 24 March 2015.

18. On the history of the library, see Bougy, *Histoire de la Bibliothèque Sainte-Geneviève*; and Zehnacker, "Bibliothèque Sainte-Geneviève."

19. Marc Smith, at the École Nationale des Chartes, Paris, evaluated the handwriting samples and communicated his opinion in an e-mail to me dated 27 May 2013.

20. Gingerich and Westman, "Wittich Connection," 28; Grafton and Weinberg, "*I Have Always Loved the Holy Tongue*," 21–22; Jardine and Grafton, "Studied for Action," 35–45, 51–55; Parenty, *Isaac Casaubon, helléniste*, 247–53.

21. Raphael, "Printing Galileo's *Discorsi*," 24.

22. See, e.g., BNCF Gal. 79, 194.

23. Mersenne, *Correspondance*, 8:94–133.

24. Gingerich and Westman, "Wittich Connection," 27–41, esp. 27, 34–35.

25. Ibid., 37.

26. On Wittich's copying of Reinhold, see ibid., 31. More generally, see Gingerich, *Book Nobody Read*, 27–41.

27. Blair, "Rise of Note-Taking in Early Modern Europe," 18.

28. Groote, Kölbl, and Weiss, "Evidence for Glarean's Music Lectures from His Students' Books," 293.

29. Descartes to Mersenne, 14 August 1634, in Descartes, *Oeuvres*, 1:304, translated in Descartes, *Philosophical Writings of Descartes*, 3:44.

30. Descartes to Mersenne, 29 June 1638, in Descartes, *Oeuvres*, 2:194, and in Mersenne, *Correspondance*, 7:314. Translated in Ariew, "Galileo in Paris," 133.

31. Ann Moss, *Printed Commonplace-Books*, 272–74. On the apparent contrast between Descartes and his bookish contemporaries, see, e.g., Grafton, "Republic of Letters in the American Colonies," 20. Richard Yeo, however, has recently provided an alternative interpretation of Descartes's position on paper technologies. According to Yeo, perhaps because Descartes placed a lesser value on the role of memory than did many of his contemporaries, he argued vigorously for the use of note-taking to make up for a poor memory or for information, such as astronomical observations, tables, and theorems, which he judged difficult to commit to memory. Yeo, "Between Memory and Paperbooks," esp. 3, 21. Descartes also took notes on his dreams and his interpretations of them, apparently adding some of these interpretations and marginalia after the fact. For an English translation of a seventeenth-century paraphrase of the notes with mention of the marginalia, see Benton, "Appendix: Descartes' *Olympica*." For the surviving fragments written by Descartes, see Descartes, *Oeuvres*, 10:179–86.

32. On Mersenne's marginalia in this copy, see Engelberg and Gertner, "Marginal Note of Mersenne"; and John Lewis, "Mersenne as Translator and Interpreter." Mersenne's annotated copy is held in the Bibliothèque des Arts et Métiers in Paris but reproduced in the 1963 facsimile. Mersenne, *Harmonie universelle*.

33. "Remarques tirées du livre de l'Harmonie universelle du P. MERSENNE." On this manuscript, see Mersenne, *Harmonie universelle*, 1:viii.

34. "Il faut supposer que la cheute des corps pesans se face dans le vuyde o dans un milieu qui n'empesche nullement, or cela posé tout corps descend aussy viste l'un que l'autre: or Galilee suppose cela dans ses derniers dialogues." BnF Fr. 12357, 5v; Mersenne, *Harmonie universelle*, 1:86.

35. In one annotation alongside a passage discussing the fall of bodies in different media, Mersenne notes his discussion of Galileo's views in the thirteenth and fourteenth articles of book 1 of his *Nouvelles pensées*: "Voyes le 13me et le 14me articles du 1er livre des nouvelles pensees de Galilee, lesquel tien dans sa response à Roca, que faisant descendre deux boules dans l'eau, l'une de . . . [illegible] et l'autre de plomb." BnF Fr. 12357, 7v; *Harmonie universelle*, 1:130. At the close of the annotation, Mersenne goes on to point out his own "experiences against Galileo," detailed at the end of article 15 of book 1. In another marginal note, Mersenne refers to both Day 1 of the *Two New Sciences* and his own discussion of the passage in his *Nouvelles pensées* about the sound waves produced by a vibrating body: "Il faut voir dans le 1er dialogue de Galileé, ou dans mon 21 article de sa version, ce qu'il . . . des ondes frémissantes d'un verre." BnF Fr. 12357, 15v; Mersenne, *Harmonie universelle*, 3:211.

36. On topical note-taking, see Blair, "Note Taking as an Art of Transmission," 86–87.

37. Jardine and Grafton, "Studied for Action," e.g., 38.

38. Müller-Wille and Scharf, "Indexing Nature," 17–18.

39. Fabbri, "Genesis of Mersenne's 'Harmonie Universelle,'" 293–94; Raphael, "Galileo's *Discorsi* and Mersenne's *Nouvelles Pensées*," 34–35.

40. For one example, consider Mersenne's discussion of Galileo's measurement of the heaviness of the air. Whereas chapters 5 and 6 indicate that university professors saw this measurement as a direct response to Aristotle's discussion of the heaviness or lightness of the elements, Mersenne noted Galileo's discussion of the issues central to the university professors' discussions very briefly (without even mentioning Aristotle by name) and focused instead on the experimental details described by Galileo. Mersenne, *Les nouvelles pensées de Galilée*, 63–64.

41. "Galilei pag. 103 de ses dialogues se trouve, car la cause porquoy le son des chordes d'or est plus bas, c'est parce qu'il est [sic] plus mou, et non parce qu'il est [sic] plus pesant, et la pesanteur d'un corps ne resiste pas davantage à la vistesse de son mouvement que la grosseur." Mersenne, *Harmonie universelle*, 3:151; transcription from BnF Fr. 12357, 15r. For Galileo's discussion, see Galilei, *EN*, 8:146.

42. Mersenne, *Correspondance*, 8:100.

43. "Le son des chordes d'or qui est plus base est à cause qu'il est [sic] plus mol, et non parce qu'il est [sic] plus pesant. Et il est faux que la pesanteur d'un corps resiste davantage à la vistesse de son movement que la grosseur." BSG, 4V 585 INV 1338 RES, 103.

44. Blair, *Too Much to Know*, 71.

45. See, e.g., Yeo, *Notebooks, English Virtuosi, and Early Modern Science*, 69–95, 175–218; and Blair, "Humanist Methods in Natural Philosophy," 541–51. Yeo has also produced a number of focused studies. See Yeo, "John Locke's 'New Method' of Commonplacing"; and Yeo, "Between Memory and Paperbooks."

46. Yale, "Marginalia, Commonplaces, and Correspondence," esp. 194, 196–97. Yale acknowledges the wider reach of these practices as they relate to collaboration and correspondence in Yale, *Sociable Knowledge*, 7–8.

47. Lenoble, *Mersenne, ou, La naissance du mécanisme*, 586–94; Taton, *Les origines de l'Académie royale des sciences*, 17–19.

48. "Il propose ce qu'il veut traiter, à sçavoir pourquoy les grandes machines, estant en tout de mesme figure et de mesme matiere que les moindres, sont plus foibles qu'elles; et pourquoy un enfant se fait moins de mal en tombant qu'un grand homme, ou un chat qu'un cheval, etc. En quoy il n'y a, ce me semble, aucune difficulté ny aucun sujet d'en faire une nouvelle science; car il est evident qu'affin que la force ou la resistence d'une grande machine soit en tout proportionnée à celle d'une petite de mesme figure, elles ne doivent pas estre de mesme matiere, mais que la grande doit estre d'une matiere d'autant plus dure, et plus malaisée à rompre, que sa figure et sa pesanteur sont plus grandes. Et il y a autant de difference entre une grande et une petite de mesme matiere, qu'entre deux egalement grandes, dont l'une est d'une matiere beaucoup moins pesante, et avec cela plus dure que l'autre." Mersenne, *Correspondance*, 8:95–96.

49. "La grande machine doit être d'un matiere d'autant plus dure e malaisée à rompre que sa figure e sa pesanteur sont plus grandes. Il y a autant de difference entre une petite et une grande machine de mesme matiere qu'entre deux egalement grandes dont l'une est d'une matiere moins pesante, et avec cela plus dure que l'autre." BSG, 4V 585 INV 1338 RES, 2.

50. "Dicimus machinam majorem confici debere quae sit eo durior et ruptu difficilior quo illius pondus et figura majora fuerint, tantumque fore discrimen inter minorem et majorem divisionem materiae, quantum fuerit inter duas aequalis magnitudinis, quarum una constat ex materia minus ponderosa et tamen duriori quam altera." Mersenne, *Correspondance*, 12:220–21.

51. Ibid., 8:100.

52. "Il . . . que ses corps ont de la pesanteur dans la vuide, ce qui n'est pas." BSG 4V 585 INV 1338 RES, 73. The ellipses in the citation indicate one illegible word; I have substituted a likely replacement in brackets in the text. Mersenne, *Correspondance*, 12:222.

53. For two examples in which the three sources are substantially the same, see the remarks pertaining to pages 50 and 54 in the 1638 printed edition: "Tout ce qu'il dit de la rarefaction et condensation n'est qu'un sophisme, car le cercle ne laisse point de parties vuides entre ses points, mais il se meut seulement plus lentement. Lorsqu'un corps se condense c'est que ses pores s'estrecissent, et qu'il en sort une partie de la matiere subtile qui les remplissoit, que l'un replit une sponge, et . . . il s'eslargisse, il y entre plus de matiere subtile." BSG, 4V 585 INV 1338 RES, 50. "Ce qu'il dit de l'or tiré n'est point a propos, car il ne se rarefie point mais il change seulement de figure." Ibid., 54. The corresponding passages in the 1638 and 1643 letters are found in Mersenne, *Correspondance*, 8:98–99 and 12:221.

54. Here the BSG copy merely states that "ces miroirs d' Archimede sont impossibles. Voyez la 119 page de la Dioptrique." BSG, 4V 585 INV 1338 RES, 42. Both Descartes's letter and the 1643 letter explicitly criticize Galileo's knowledge of catoptrics. Descartes's letter states, "Il monstre n'estre pas sçavant en la catoptrique, de croire ce qui se dit des miroirs ardans d'Archimede, lesquels j'ay demonstré estre impossibles en ma *Diop.*, p. 119." Mersenne, *Correspondance*, 8:98. The 1643 letter states, "Dicimus specula illa Archimedea, pag. 42, esse impossibilia, et in eo Cavalerium ignoratione verae catoptricae errasse, cujus rei rationem explicabimus, si quis ea de re laboret." Ibid., 12:221.

55. "Si quae sint vera, proponere . . . [illegible] verum, sed ad . . . [illegible] verum non probat." BSG, 4V 585 INV 1338 RES, 288.

56. "Si quae ait pag. 236 à sectione illa (mobile quoddam super planum et caet.), vera non sunt, corruit totus tractatus. Atqui vera esse non probat." Mersenne, *Correspondance*, 12:222.

57. "Il adjouste une autre supposition aux precedentes, laquelle n'est pas plus vraye, à sçavoir que les cors jetez en l'air vont egalement viste suivant l'horizon; mais qu'en descendant leur vitesse s'augmente en proportion double de l'espace. Or cela posé, il est tres aisé de conclure que le mouvement des cors jetez devroit suivre une ligne parabolique; mais ses positions estant fausses, sa conclusion peut bien aussy estre fort esloignée de la verité." Ibid., 8:102. Translation from Descartes, *Philosophical Writings of Descartes*, 3:126.

58. One additional remark that does not reference a specific page appears to correspond to a comment found in Descartes's letter and in the BSG copy. In his letter to Mersenne, Descartes wrote, "Il compare la force qu'il faut pour rompre un baston de travers, avec celle qu'il faut pour le rompre en le tirant de haut en bas, et dit que, de travers, c'est comme un levier dont le soustien est au milieu de son espaisseur: ce qui n'est

nullement vray, et il n'en donne aucune preuve." Mersenne, *Correspondance*, 8:100. Mersenne transcribed the annotation as, "Il ne preuve point que de travers le baston soit comme un levier, dont le soustien est au milieu de son espaisseur, ce qui n'est pas vray." BSG, 4V 585 INV 1338 RES, 114. In the 1643 letter, Mersenne wrote, "Non probat transversum prisma esse instar vectis cujus hypomochlion sit in medio densitatis, neque id verum est." Mersenne, *Correspondance*, 12:222.

59. These two remarks referring to experiments are: "Quod ait gravia descendentia eo tardius in aqua, aere et caet. descendere, quam in vacuo, quo plus de pondere illorum tollitur ab aere, aqua vel alio medio, repugnant experientiae" and "Adversus ea quae pag. 280 dicit de amplitudinibus semiparabolarum et caet. Dicimus haec omnia repugnare experientiae." Mersenne, *Correspondance*, 12:223, 222–23. This first remark refers to page 69 of the 1638 edition. Descartes had mentioned this page in his 1638 letter and Mersenne had annotated it, but their remarks do not correspond to that found in the 1643 letter.

60. BSG, 4V 585 INV 1338 RES, 22, 28, 31, 40, 42, 43, 48, 50.

61. "Ces miroirs d'Archimede sont impossibles. Voyez la 119 page de la Dioptrique." Ibid., 42.

62. Mersenne, *Les nouvelles pensées de Galilée*, 27.

63. Ibid., 28.

64. Ibid.

65. Mersenne, Ibid., i–ix.

66. Mersenne, *Les méchaniques de Galilée*, 17–20.

67. Drabkin, "Aristotle's Wheel"; Palmerino, "Galileo's and Gassendi's Solutions."

68. "Tout ce qu'il dit de la rarefaction et condensation n'est qu'un sophisme, car le cercle ne laisse point de parties vuides entre ses points, mais il se meut seulement plus lentement. Lorsqu'un corps se condense c'est que ses pores s'estrecissent, et qu'il en sort une partie de la matiere subtile qui les remplissoit . . . et quand il s'eslargisse, il y entre plus de matiere subtile." BSG, 4V 585 INV 1338 RES, 50.

69. Mersenne, *Les nouvelles pensées de Galilée*, 31–32.

70. Ibid., 28.

71. Descartes, *Oeuvres*, 6:84, translation from Descartes, *Discourse on Method, Optics, Geometry, and Meteorology*, 67.

72. Mersenne, *Correspondance*, 12:221. These criticisms were directed at pages 28, 42, and 50 of the 1638 text.

73. Descartes, *Oeuvres*, 1:309–12.

74. See, e.g., the letters of 23 November 1625 and 30 April 1630 in Mersenne, *Correspondance*, 1:308–11 and 2:456.

75. Ibid., 7:232.

76. See, e.g., ibid., 4:212; 5:238; 6:167–77.

77. Mersenne, *Les méchaniques de Galilée*, 18–19.

78. Mersenne, *Correspondance*, 7:177.

79. Blair, *Theater of Nature*, 200–201.

80. Sherman, *Used Books*, 12–13.

81. Renn and Damerow, *Equilibrium Controversy*, 178, 181, 184.

82. Ibid., 203.

83. Ibid., 203–5, 248.

84. Dear, *Mersenne and the Learning of the Schools*, esp. 9–22.

Chapter 4: An Annotated Book of Many Uses

1. Bennett, "Hooke and Wren and the System of the World," 38; Bennett, *Mathematical Science of Christopher Wren*, 61–62.

2. On the discovery of these copies and key features, see Raphael, "Galileo's *Discorsi* as a Tool for the Analytical Art"; and Raphael, "Reading Galileo in the Early Modern University."

3. Bennett, *Mathematical Science of Christopher Wren*, 29–32, 38–43, 71–73.

4. "Aliam . . . demonstrationem, vide in pag. vacua ad initium libri." Bodleian, Savile Bb.13, 118.

5. "Vide Demonstrationem aliam D. Wardi." Bodleian, Savile A.19, 89.

6. Bennett, *Mathematical Science of Christopher Wren*, 14–25.

7. On declarations of this trifold relationship in the eighteenth century and links between the Accademia del Cimento and the Royal Society, see Feingold, "Accademia del Cimento and the Royal Society."

8. Fiocca, "Galileiani e Gesuiti a Ferrara nel Seicento"; Gavagna, "I gesuiti e la polemica sul vuoto"; Gavagna, "Paolo Casati e la scuola galileiana."

9. Vilain, "Christiann Huygens' Galilean Mechanics."

10. "Supponit hoc experimentum partes axis IK arctius sibi invicem cohaerere, quam superficiem cylindri EF, superficiei aquae sibi contiguam. Alias disrumpetur axis IK, manente cylindro EH ab aquae superficie, indivulso." Bodleian, Savile Bb.13, 15, and Savile A.19, 11.

11. "Si divulsa Cylindri EH superficie EF, ab aquae superficie sibi contigua, axis IK (cuius diameter longe minor est diametro Cylindri GF, vel aquae) momento ponderis appensi non dirumperetur; manifestum est, absque ulteriori experimento, aliam esse causam cohaerentiae partium in axe ferreo, quam evitationibus vacuis." Ibid. Savile A.19 has "disrumperetur" instead of "dirumperetur."

12. "ἐπέχω. Non enim repugnat aliam esse causam cohaerentiae partium continui, praeter rationem vacui, cuius tamen vis aequalis sit ipsi vacuo; in eo saltem corpore in quo periculum factum est. Pluribus igitur experimentis opus est, ad veritatem certo emendam." Bodleian, Savile Bb.13, 16, and Savile A.19, 12.

13. Robinson, "Unpublished Letter of Dr Seth Ward," 69.

14. Ibid. The significance of this quotation for Ward's note-taking practice was cited previously by Yeo. Whereas Yeo focused on the relationship between memory, note-taking, and natural history, my contribution is to remark on the particular relationship between this note-taking and experimenting in relation to a text of mechanics and mathematics like the *Two New Sciences*. See Yeo, "Between Memory and Paperbooks," 14.

15. Philip, *Bodleian Library in the Seventeenth and Eighteenth Centuries*, 1–22, 35–36, 44–47.

16. Robinson, "Unpublished Letter of Dr Seth Ward," 69. On the origin and development of such desiderata lists as advancing a communal research agenda, see Keller, "Accounting for Invention"; Keller, *Knowledge and the Public Interest*; and Keller, "New World of Sciences."

17. Frank, *Harvey and the Oxford Physiologists*, 43–89; Webster, *Great Instauration*, 144–78.

18. On these projects at Oxford, see Rhodri Lewis, *Language, Mind, and Nature*, 64–109; and Poole, *World Makers*.

19. On the religious and intellectual inspirations for new models of knowledge production, see Picciotto, *Labors of Innocence in Early Modern England*; Picciotto, "Scientific Investigations"; and Webster, *Great Instauration*.

20. For this culture of Baconian note-taking and Ward's membership in it, see Yeo, "Between Memory and Paperbooks," 3–4.

21. Hunter, *Boyle Papers*, 37–44; Yeo, *Notebooks, English Virtuosi, and Early Modern Science*, 151–73. On the cultural and social aspirations of this milieu, see Shapin, *Social History of Truth*, esp. 126–92.

22. Johns, "Reading and Experiment in the Early Royal Society"; Yeo, *Notebooks, English Virtuosi, and Early Modern Science*, 224–30.

23. "Modus conficiendi filamenta aurea." Bodleian, Savile Bb.13, 53; Savile A.19, 40. "Modus accuratissimus librandi aquas." Savile Bb.13, 71; Savile A.19, 53. "Modus investigandi rationem velocitatum gravium in eodem vel diversis mediis." Savile Bb.13, 76; Savile A.19, 57. "Modus investigandi gravitatem aeris." Savile Bb.13, 79; Savile A.19, 60. "Modi tres accendi tonum chordae." Savile Bb.13, 100; Savile A.19, 75. "Modi ducendi lineam parabolicam." Savile Bb.13, 146; Savile A.19, 110. "Modus mensurandi tempus." Savile Bb.13, 176; Savile A.19, 132.

24. Bodleian, Savile Aa.12. Shelfmarks Savile N–Z, Aa, and Bb correspond to Savile's original donations, but these shelfmarks also include later additions. Craster, *History of the Bodleian Library*, 185.

25. "Filtratio." Bodleian, Savile Aa.12, 2.18. "Distillatio." Savile Aa.12, 2.19.

26. "Modus eliciendi aquas et extracta rebus succosis." Bodleian, Savile Aa.12, 3.8.

27. In the word *modus*, one reader of the copy consistently included ligatures between the *m* and the *o* and between the *u* and the *s*, but he did not connect the *d* and the *u*. Other passages in the book, in contrast, show ligatures between the *d*, the *u*, and the *s* but not between the *m* and the *o*.

28. Webster, *Great Instauration*, 164–72.

29. Ward and Wilkins, *Vindiciae Academiarum*, 36.

30. Feingold, "Origins of the Royal Society Revisited"; Hunter, *Establishing the New Science*, 36; Lisa Jardine, *On a Grander Scale*, 111–15; Malcolm, "Hobbes and the Royal Society"; Poole, *World Makers*, xiv.

31. "Modus accuratissimus librandi aquas." Bodleian, Savile Bb.13, 71, and Savile A.19, 53; Galilei, *EN*, 8:114.

32. See, e.g., Birch, *History of the Royal Society*, 1:459.

33. "Modus mensurandi tempus." Bodleian, Savile Bb.13, 176, and Savile A.19, 132; Galilei, *EN*, 8:213.

34. Birch, *History of the Royal Society*, 1:461, 464, 466, 468–69, 471, and 3:398; Lisa Jardine, "Monuments and Microscopes."

35. Birch, *History of the Royal Society*, 1:461.

36. "Modus investigandi rationem velocitatum gravium in eodem vel diversis mediis." Bodleian, Savile Bb.13, 76; Savile A.19, 57.

37. "Velocitatem descensus non pendere a gravitate mobilis." Bodleian, Savile Bb.13, 74; Savile A.19, 56. "Idem medium magis resistit velociori motui." Savile Bb.13, 75; Savile A.19, 56.

38. "Esto gravitas aëris=1.

plumbi=10000.

ebeni=1000.

Posita absoluta gravitate plumbi et ebeni (hoc est, in medio cui nulla inest resistentia, ut in vacuo) inter se aequali; resistentia aeris detrahet gravitati plumbi (10000) gradum unum: gravitati autem ebeni (10000) gradus 10. Ergo,

Gravitas plumbi in aere erit 9999.

—Ebeni in aere erit 9990.

Ergo velocitas plumbi in aere erit grad. 9999

—Ebeni in aere, grad. 9990

Divisa igitur altitudine qualis in 10,000 partes ut demissis per aera aequalibus plumbi et ebeni figuris: cum plumbum appulerit ad gradum ultimum, vel ad 9999; ebenum erit solum in gradu 9990; scilicet 9 gradus post plumbum.

Plumbum est duodecis gravius aqua. Ebur bis gravius aqua.

Posita igitur gravitate plumbi et eboris absoluta aequali, aqua velocitati plumbi tollet partem duodecimam; velocitati eboris dimidium. In descensu igitur plumbi et eboris in aqua, cum plumbum fuerit in fundo, ebur erit solum ad partem dimidiam totius altitudinis."

Bodleian, Savile Bb.13, loose sheet verso between 18 and 19; Savile A.19, loose slip 1v between 88 and 89.

39. Birch, *History of the Royal Society,* 1:465.

40. Ibid., 1:469.

41. Galilei, *EN,* 8:107–9.

42. Birch, *History of the Royal Society,* 1:475–76.

43. Birch wrote that it was the reading of a treatise by Mersenne and the consultation of Galileo's writing, ostensibly the *Two New Sciences,* that prompted members to engage in experimental investigations into the resistance of materials to breaking. Ibid., 1:109. An account of the experiments is found at ibid., 1:109, 113, 401–2, 405, 409, 415, 456.

44. Schemmel, *English Galileo,* 1:47.

45. BNCF Gal. 80, 194.

46. See, e.g., Bodleian, Savile X.9(2), 11r.

47. The problems to which Ward offers an analytical solution are found in Galilei, *EN,* 8:89–90, 168–69, 177, 232. See also Raphael, "Galileo's *Discorsi* as a Tool for the Analytical Art."

48. On Viète and his introduction in England, see Stedall, *Discourse concerning Algebra,* 55–87.

49. Ibid., 77–82. On the influence on Wallis of Descartes and Torricelli, see Stedall, *Arithmetic of Infinitesimals,* xii–xvii.

50. Sherman, *Used Books,* 13–15.

51. Raphael, "Galileo's *Discorsi* as a Tool for the Analytical Art," 119.

52. For an overview of this reception of Newton's work, see Bertoloni Meli, *Thinking with Objects,* 302–7.

53. For examples of these views, see Blay, *Reasoning with the Infinite*, 1–12; and Thomas L. Hankins, *Science and the Enlightenment*, 17–23.

54. "Resistentia vacui," "De Indivisibilibus Infinitis," "Exempla Rarefactionis et Condensationis." Bodleian, Savile Bb.13, 12, 25, 61; Savile A.19, 9, 19, 46.

55. "Continuum constare ex Atomis Indivisibilibus." Bodleian, Savile Bb.13, 49; Savile A.19, 37. "Velocitatem descensus non pendere a gravitate mobilis." Savile Bb.13, 74; Savile A.19, 56. "Aer gravis." Ward only, Savile Bb.13, 78. "Grave in motu differt a Gravi in quiete." Savile Bb.13, 64; Savile A.19, 48.

56. On the standardized nature of university teaching and the continued relevance of the Aristotelian framework in university pedagogy in the seventeenth century, see, e.g., Baldini, "Development of Jesuit 'Physics' in Italy," 252; Brockliss, "Curricula," 580; Brockliss, *French Higher Education in the Seventeenth and Eighteenth Centuries*, 337; and Grendler, *Universities of the Italian Renaissance*, 277.

57. Sherman, *John Dee*, 72–73.

58. See the description of the extant copy, Naples, Biblioteca Nazionale "Vittorio Emanuele III," shelfmark S.Q. XXXII.B.90, in Westman, "Reception of Galileo's 'Dialogue,'" 344.

59. Blair, *Too Much to Know*, 71.

60. "Condensatio et Rarefactio ex sententia Galilei. Omnia corpora constant ex Indivisibilibus infinitis, partim plenis, partim vacuis: quorum compressione, fit condensatio; dilatione, Rarefactio: Alias, vel Penetratio fieret quantorum, vel vacuum quantum." Bodleian, Savile Bb.13, inside front cover; Savile A.19, loose slip 4r between 88 and 89.

61. "Causa accelerationis motus in gravium descensu." Bodleian, Savile Bb.13, 161; Savile A.19, 122.

62. While the *Two New Sciences* is widely celebrated for its study of the quantitative aspects of motion without recourse to a search for causes, historians have acknowledged that Galileo was not completely uninterested in explaining causes. See, e.g., Garber, "On the Frontlines of the Scientific Revolution," 148–51; and Bertoloni Meli, "Mechanics," 662.

63. "Si vera esset hypothesis haec Platonica, oportuit terram caeterosque planetas centrum aliquid coelestem (solem fortasse) in motu eorum recto respexisse (sicut gravia nostra ad terram centrum feruntur) aut alia quaedam assignanda est causa motus accelerati." Bodleian, Savile Bb.13, 254; Savile A.19, 193.

64. "Speculum ustorium Bonaventuri Cavalieris." Ward only, Bodleian, Savile Bb.13, 42. "Vid. Clavium in Scalig. Cyclometr." Savile Bb.13, 48; Savile A.19, 36 ("adv" in place of "in"). "Clavium videtur intelligere." Ward only, Savile Bb.13, 58. "Galilei tract. de Aqua nulla ei tenacitate inesse." Savile Bb.13, 71; Savile A.19, 53.

65. "Aristotelis error." Bodleian, Savile Bb.13, 65; Savile A.19, 49.

66. "Est Aristotelis propria." Bodleian, Savile A.19, 15. Galileo's argument and its relationship to Aristotle's assertion of nature's abhorrence of the void are addressed in Galilei, *Two New Sciences*, 27n8.

67. Gingerich and Westman, "Wittich Connection," 28, 34.

68. Renn and Damerow, *Equilibrium Controversy*, 182, 238–40.

69. Blair, *Too Much to Know*, 133–35.

70. Blair, *Theater of Nature*, 197–98.

71. Clark, "Dividing Time," 116–17.

72. Ibid., 116–19.

73. Gingerich and Westman, "Wittich Connection," 28.

74. Bennett, "Hooke and Wren and the System of the World," 38.

75. Blair, "Note Taking as an Art of Transmission"; Blair, "Rise of Note-Taking in Early Modern Europe"; Grafton and Jardine, *From Humanism to the Humanities*, esp. 9–22.

76. Translation and quotation from Grendler, *Schooling in Renaissance Italy*, 245.

77. Grafton, "Teacher, Text and Classroom," 41–42.

78. For an example of the copying of astronomical models not included in Copernicus's original, see Gingerich and Westman, "Wittich Connection," 31.

79. Ibid., 37.

80. Groote, Kölbl, and Weiss, "Evidence for Glarean's Music Lectures from His Students' Books," 281.

81. Oosterhoff, "Book, a Pen, and the Sphere."

82. Goulding, "Polemic in the Margin"; Stedall, *Discourse concerning Algebra*, 77–82.

83. For some examples, see Bodleian, Savile B.14 (Hevelius's *Prodromus Cometicus*, *Descriptio Cometae Anno . . . mclclxv exorti*, and *Cometographia*), Savile C.1 and Savile C.2 (Cardano's *Opera Omnia*), and Savile K.3 (Giovanni Alfonso Borelli's *Euclides Restitutus*).

84. Feingold, "Galileo in England," 414–16, 419.

85. On the Allestree Library, see Neagu, "Time Capsule under Restoration." The copy of the *Two New Sciences* is Christ Church Library, Allestree c.5.21.

86. Blair, "Rise of Note-Taking in Early Modern Europe," 313. Blair's account draws on Hess, "Fundamenta fürstlicher Tugend," 131–74.

87. Wolfenbüttel, Herzog-August-Bibliothek A: 33.9 Quod. Information on the copy provided by Christian Hogrefe in an e-mail to the author on 21 November 2012.

88. Bibliothèque interuniversitaire de la Sorbonne, RR 8=89.

89. Hunt and Watson, *Bodleian Library Quarto Catalogues. IX. Digby Manuscripts*.

90. Their copies are, respectively, Greifswald, Universitätsbibliothek, 520/Rb 75(2), and Trinity College Library, Dublin, L.ee.11.

91. Needham, *Galileo Makes a Book*, 49–50. For the presence and absence of annotations in surviving copies, see ibid., 2:213–35.

92. Westman, "Reception of Galileo's 'Dialogue,'" esp. 334.

93. Gingerich and Westman, "Wittich Connection," 28; Grafton and Weinberg, "*I Have Always Loved the Holy Tongue*," 21–22; Jardine and Grafton, "Studied for Action," 35–45, 51–55; Parenty, *Isaac Casaubon, helléniste*, 247–53. Note, however, that not all sixteenth-century annotators left statements of their intent. See, e.g., the example of Paul Wittich in Gingerich and Westman, "Wittich Connection," 33.

94. Yeo, "Loose Notes and Capacious Memory"; Yeo, *Notebooks, English Virtuosi, and Early Modern Science*, 151–73.

95. Harvey's Livy, for example, contained Simon Grynaeus's instructions for reading history. Jardine and Grafton, "Studied for Action," 36.

96. Raphael, "Galileo's *Two New Sciences* as a Model of Reading Practice."

97. For a well-known example, see Jardine and Grafton, "Studied for Action," esp. 30–33.

Chapter 5: The University of Pisa and a Dialogue between Old and New

1. Grendler, *Universities of the Italian Renaissance*, 477–83.

2. Consider, for example, the response of the Italian university professors Giovanni Antonio Magini, at Bologna, and Claude Bérigard and Scipione Chiaramonti, at Pisa, to Galileo's telescopic discoveries. Galilei, *Sidereus Nuncius or The Sidereal Messenger*, 87–111; Stabile, "Il primo oppositore del *Dialogo*." On Galileo and natural philosophers at Pisa, see, e.g., Biagioli, *Galileo, Courtier*, 159–244.

3. For examples, see Gomez Lopez, "Donato Rossetti et le Cercle pisan"; Gomez Lopez, *Le passioni degli atomi*; and Gomez Lopez, "Dopo Borelli."

4. Baldini, "Tra due paradigmi?," 189–90.

5. On the Cimento, see Middleton, *Experimenters*; and Beretta, Clericuzio, and Principe, *Accademia del Cimento and its European Context*.

6. Gomez Lopez, "Dopo Borelli," 228–32; Galluzzi, "La scienza davanti alla Chiesa e al Principe."

7. On this difference, see Baroncini, *Forme di esperienza e rivoluzione scientifica*, esp. 39–62; Daston, "Baconian Facts, Academic Civility, and the Prehistory of Objectivity"; Gaukroger, *Explanatory Structures*; Schmitt, "Experience and Experiment"; and Dear, *Discipline and Experience*, 21–25.

8. Maffei, "Relazione di Giovanni Maffei," transcription in Galluzzi, "La scienza davanti alla Chiesa e al Principe," 1329.

9. Ibid.

10. Marchetti, "Lettera di Alessandro Marchetti a Leopoldo dei Medici," transcription in Galluzzi, "La scienza davanti alla Chiesa e al Principe," 1334–35.

11. "Minuta autografa di Lorenzo Bellini a Domenico Magni," transcription in Galluzzi, "La scienza davanti alla Chiesa e al Principe," 1331.

12. "Lettera di G. A. Borelli a Leopoldo dei Medici," transcription in Galluzzi, "La scienza davanti alla Chiesa e al Principe," 1333.

13. For this hypothesis, see Stabile, "Il primo oppositore del *Dialogo*," esp. 278–79. On Bérigard more generally, see Bellucci, "La filosofia naturale di Claudio Berigardo"; Favaro, "Oppositori di Galileo: IV. Claudio Berigardo"; "Beauregard, Claudio Guillermet"; and Stabile, *Claude Bérigard, 1592–1663*.

14. See, e.g., Bérigard, *Circulus Pisanus* (1643), pt. 1, dedicatory letter, unpaginated.

15. BML, Bellini, "Opere Varie," 1:200v and 4:173r, 183r.

16. Bérigard, *Circulus Pisanus* (1643), 2:2–3. On these names, see Iofrida, "La filosofia e la medicina," 306.

17. Bérigard, *Circulus Pisanus* (1643), 3:35.

18. Ibid., 3:41.

19. Drabkin, "Aristotle's Wheel"; Palmerino, "Galileo's and Gassendi's Solutions." See the earlier discussion of this paradox in chapter 3.

20. Galilei, *EN*, 8:68–71.

21. Ibid., 8:93–96.

22. Schmitt, "Eclectic Aristotelianism."

23. For Galileo's and his readers' (including Bérigard's) use of diagrams, see Raphael, "Teaching through Diagrams."

24. Bérigard, *Circulus Pisanus* (1643), 3:32.

25. Ibid.

26. Ibid., 3:32–33.

27. Galilei, *EN*, 8:62–63.

28. Ibid., 8:60.

29. Bérigard, *Circulus Pisanus* (1661), 1:55. For his substitution of Galileo's device in the section of *On Generation*, see ibid., 1:390.

30. The narrative that follows is developed from Dear, *Discipline and Experience*, 180–209; Hellyer, *Catholic Physics*, 142–58; and Shea, *Designing Experiments and Games of Chance*, 17–185. See also Middleton, *Experimenters*; Montacutelli, "Air 'Particulae' and Mechanical Motions"; and Shapin and Schaffer, *Leviathan and the Air-Pump*.

31. For Kircher's experiment, see Kircher, *Musurgia Universalis*, 1:11–13; and Shea, *Designing Experiments and Games of Chance*, 28.

32. Bérigard, *Circulus Pisanus* (1661), 1:55.

33. Shea, *Designing Experiments and Games of Chance*, 32–36.

34. Bérigard, *Circulus Pisanus* (1647), 137.

35. French, *William Harvey's Natural Philosophy*, 246–51. French describes Bérigard's text similarly to the way I do, though he attributes a different motive to Bérigard, arguing that Bérigard's treatment of Harvey was "a very small part of his effort to make sure that a Counter-Reformation Aristotle could meet all demands made by neoteric philosophy" (251).

36. Brockliss, "Curricula," 582–83.

37. Palmerino, "La fortuna della scienza galileiana," 75–78. With some exceptions, the authors of this contribution and those cited below do not indicate whether their actors relied on the *Dialogue*, the *Two New Sciences*, or another text when discussing Galileo.

38. Baldini, "Tra due paradigmi?," 215.

39. Navarro Brotons, "Filosofia natural y disciplinas matematicas," 97–102.

40. Schmitt, "Galilei and the Seventeenth-Century Text-Book Tradition," 225.

41. Baldini, "Tra due paradigmi?," 217n101.

42. Ibid., 215.

43. Ibid., 191.

44. Galilei, *EN*, 20:517–18.

45. These treatises are found in BNCF Gal. 114, 1r–29v ("De Triangulorum Analysis"), 30r–32v ("Fragmentum Geometricum"), 33r–77r ("Scritti relative alla geometria pratica"), 85r ("Praelectio in Librum Quintum Euclidis"), 79r–83r ("Praelectio"), 103r–105v ("Fortificazioni"); and Gal. 115, 1r–82v ("De Sphaera Mundi"), 83r–106v ("Frammenti autografi di un trattato latino di gnomonica").

46. For example, Galileo showed how to solve similar problems in his 1606 *Le operazioni del compasso geometrico et militare*. See Galilei, *EN*, 2:414–23.

47. "Misurare un altezza benche non si possa uedere per mezzo d'un filo che cadendo dalla sommita d'essa solo nell'estremità inferiore si ueda." Renieri, BNCF Gal. 114, 46v.

48. "Dipende questo nobilissimo Problema da una gentile osservazione del Signore Galilei Galilei [*sic*] fatta intorno alle uibrazioni che fanno i pesi a qualche filo attaccati qualunque uolta siano mossi dal perpendicolo." Ibid.

49. "Ciò dunque supposto sia AB un filo pendente dalla sommità della torre FAE et in B posto un graue uadia uibrandosi in F et E ne possa l'osseruatore uedere altro di detto filo che la parte GB come auuerebbe [*sic*] se forse dentro la porta della torre per sapere adunque quanta sia l'altezza BA che non si uede prendasi un filo minore CD col graue in D attaccato e nello stesso tempo che si contano le uibrazioni di B in F et E si contino le uibrazioni di D in H et I imperoché se piglierimo la proporzione duplicata del numero delle uibrazioni D al numero delle uibrazioni B la stessa sarà quella della lunghezza AB alla lunghezza CD; perche dunque i numeri quadrati sono in duplicata proporzione delle loro radici se moltiplicheremo in se stesso sí il numero delle uibrazioni D come anco quello di B il prodotto di D al prodotto di B haurà la proporzione di AB a CD per tanto pongasi che nello stesso tempo che si sono contate 20 uibrazioni del pendulo B nello stesso se ne siano numerate 240 del pendulo D e moltiplichisi in se stesso 20 e poi 240 sí che 400 sia quello e 57600 sia quest'altro diró adunque che l'altezza AB, contiene 57600 di quelle parti delle quali CD ne contiene 400; posto adunque che CD si fusse preso di lunghezza d'un braccio diuidendo 57600 per 400, ne risulterebbe la lunghezza AB di braccia 144 perché se CD di 400 da BA di 57600 lo stesso CD di 1 dara BA di 144." Ibid., 46v–47r.

50. Galilei, *EN*, 8:140.

51. "Sexto fluxum atque refluxum maris si terreno motui non attribuatur difficilime alteri causae assignare posse contendunt." Renieri, BNCF Gal. 115, 32r.

52. "In maiori alicuius navis cubiculo quae velocissime . . . sintque in eo muscae . . . habeatque insuper vas aqua plenum per quod pisciculi vagari possint." Ibid., 33r.

53. Fabroni, *Historiae Academiae Pisanae*, 3:410–13. On Giannetti, see Preti, "Giannetti, Pascasio," and the sources cited therein.

54. "Physica Excellentissimi Viri Pasqualis Giannetti in Pisano Lycaeo Publici Medicinae ac Philosophiae Professoris." BML, Giannetti, "Physica," title page.

55. BML, Giannetti, "Physica," 6v.

56. "Eam Philosophiae partem aggredimur, quae Physica, seu naturalis appellatur. Quum autem variae, atque pugnantes inter se Philosophorum sententiae, hanc scientiam perdifficilem, atque adeo perplexam effecerunt, ut vel placita illorum referre, et expendere sit laboris immensi, praecipue ex iis, et maxime utilia seligemus, brevem simul, et facilem philosophandi rationem seligentes." Ibid., 2r–v.

57. "De Motu Proiectorum," "De Radio Refracto." Ibid., 38r, 48v.

58. "Id ergo semper fuit observatum, et vulgatissimum corpora gravia deorsum cadentia sensim fieri velociora; sed qua ratione crescat velocitas, et crescant spatia temporibus aequalibus peragrata nemo ante Galilaeum ostendit. Nos eo ordino progredientes, quem rei explicandae censemus aptissimum illud sumimus iam exploratum experimentis innumeris, et ab omnibus receptum, spatia peracta a corpore gravi ex quieto descendente, si numerentur unum post aliud crescere secundum rationem numerorum imparium ab unitate, videlicet 1, 3, 5, 7, 9." Ibid., 32v.

59. "Hinc sequitur, quod si huiusmodi spatia comparentur inter se singula incoata a quiete habent inter se rationem duplicatam temporum, seu sunt inter se ut temporum quadrata, quod nullam habet difficultatem, et ex sola vocum, seu nominum explicatione

intelligitur, cuiuslibet enim numeri quadratum vocatur numerus ille, qui conflatur ex multiplicatione illius in se ipsum, ut quadratum unitatis est comitas ipsa, quadratum II, est IV, quadratum ternarii est IX etc. Quum ergo spatia crescant consequenter numerata secundum numeros impares, si ex quiete, seu initio motus sumpta conferantur inter se erunt, ut temporum quadrata. Patet id in linea exposita A etc., nam quum spatium BC triplum sit spatii AB, erit totum CA eiusdem AB quadruplum: est autem CA spatium peractum intra duo scrupula, igitur eam habet rationem ad AB, quam habet quadratum II, ad I. Spatium autem CB tertio scrupulo peractum positum fuit eiusdem AB quintuplum; erit igitur totum spatium AD, eiusdem AB noniplum, nempe ut quadratum numeris III, quod est IX, ad I." Ibid., 33r.

60. "Hic obiter monere volumus, errorem inesse in Dialogo Secundo Galilaei de Mundi Sistemate, ubi habet pilam ferream centum librum cadere ex altitudine Bracchiorum 100 exactis 5 minutis secundis, quum iuxta receptissimam, et certissimis experimentis probatam sententiam spatium a pila ferrea intra 5 minuta transactum sit plusquam duplum; excedat scilicet Brachia 1200." Ibid., 33v.

61. Bertoloni Meli, *Thinking with Objects*, 108, 110–11.

62. "Haec proportio inter spatia peragrata, servatur etiam, quum grave cadit per planum inclinatum ad Horizontem, quamvis descendat minus velociter, quam si cadat in perpendiculo, atque pariter si idem grave fuerit appensum filis diversae longitudinis, et remo tum a perpendiculo sub aequalibus angulis, observantur spatia peragrata ex quiete, licet sint circulorum circumferentiae, eam inter se rationem servare, quam exposuimus, scilicet quam habent temporum quadrata." BML, Giannetti, "Physica," 33v–34r.

63. "Et exordiamur a motu, quo omnium minime violentus videtur eum proponimus, qui imprimitur a motu translato, ut e.g. quem simul cum navi suscipiunt omnia corpora quae in illa continentur, et simul transferuntur corpora haec eodem affici motu, atque impetu, quo afficitur navis translata satis constat, et nos pluries experti sumus hoc ipsum, quando cimba recti sumus; si enim illa ad ripam impegit, sensimus impetum impressum, ex quo nisi quis caveat, cadit anteorsum, si cimba repente sistatur, atque quum diutius in ea vel sedentes moramur, postea sentimus corporis lassitudinem ortam ex eo, quod ut corpus ipsum libratum detineamus, dum variis successionibus cima impellitur, cogimus musculos dorsi identidem contrahere, atque lassare, ne anteorsum, vel retrorsum cadamus. Sed evidenter impressi motus testimonium praebent gravia, quae dum navis veloci cursu transfertur demittuntur ex summitate mali ea enim ita descendunt, ut non magis ab ipso removeantur, quam quum navis omnino quiescit, neque tempus descensus longius, aut brevius apparet." Ibid., 38v–39r. Note that Giannetti describes the example using the first-person plural, a grammatical sign that he intends this example to be an experience common to all and thus, according to traditional natural philosophical standards, suitable evidence in support of his conclusions. On the use of "common" experiences as elements of natural philosophical proofs, see Dear, *Discipline and Experience*, esp. 21–25, 63–92.

64. "Refert Gassendus rem primum a Galilaeo propositam a plerisque incredibilem putatam fuisse, atque experimentum fuisse propterea institutum in triremi egregie instructa, cuius ea fuit velocitas, ut intra horae quadrantem pervaderet quatuor miliaria."

BML, Giannetti, "Physica," 39r. On Gassendi's experiment, see Gassendi, *Opera Omnia* 3: esp. 478A–B, 481B–483A.

65. "Sed dehinc videantur Galilaeus, et Torricellius." BML, Giannetti, "Physica," 40r.

66. "Si ergo aliquis velit regula, et circino hanc lineam describere, ducat ex dato puncto duas rectas lineas rectum angulum comprehendentes, et unam dividat in quot libuerit particulas ita crescentes, ut crescunt spatia a gravi cadente peracta: aliam vero in totidem partes aequales dividat, et a punctis sectionum utriusque lineas alias ducat prioribus parallelas, quousque se mutuo secent. Linea curva ab angulo ducta per puncta sectionum huiusmodi linearum representabit exacte parabolam; sic enim vocant hanc lineam, per quam omnia missilia seu proiecta gravia oblique impulsa feruntur." Ibid., 40r.

67. On the use of diagrams in classroom teaching of mathematics and natural philosophy, see Raphael, "Teaching through Diagrams."

68. "Hinc conspecta est . . . [illegible] ad militares usus maxime utilis, et confectae tabulae, quarum beneficio cognita distantia, ad quam excutitur globus a tormento quodam, facile colligitur, quae nam futura sit distantia, ad quam pertinget Globis ipse explosus in quacumque tormenti positura, sive, ut aiunt, elevatione, variae enim sunt distantiae huiusmodi, et quidem illa maxima est, quae consequitur elevationem tormenti mediam, id est, quum axis ipsius constituit cum linea horizontali angulum semirectum." BML, Giannetti, "Physica," 40r.

69. "Huius Globi materiam videmus in quasdam massas discretas, quae vulgo vocantur elementa, et sunt Terra, Aqua, Aer, et Ignis. Videmus haec motibus diversis agitari, misceri diversimode, cohaerere, atque dissolvi, modo sursum, modo deorsum moveri. Huius tumultus originem, et causam ante omnia oportet exquirere, quum hinc plurima pendeant, in quibus explicandis laborant Philosophi. Si materia, ex qua haec elementa constant creata fuisset absque motu seu energia, ac propensione ad motum, iacuisset perpetuum iners rudis indigestaque moles, quocirca quum in eum finem materia sit procreata, ut ex ea omnis generis corpora efficiantur, necesse est in ipsa agnoscere insitam initio vim quamdam motricem. Aristoteles, et eius sectatores vim huiusmodi duplicem agnoverunt ad terminos plane diversos, et oppositos trahentem elementa, videlicet quam dixere gravitatem semper trahentem demum, et dirigentem ad Terrae antrum corpora, quibus inest, alteram vero voluerunt esse vocatam levitatem, ducentem corpora sursum." Ibid., 4v–5r.

70. "Duplex hoc motus incitamentum nulla ratione ostenditur, immo vero cum ratione videtur pugnare: nam globum hunc voluit natura procul dubio conservari, levitas autem partes incitat ad discessum, ideoque dissolutionem molitur. Gravitatem, seu conatum descendendi res ipsa demonstrat, sed levitatem nonnisi ex corporum quorumdam ascensu deducunt, atque si ea ascenderent in spatio prorsus vacuo, et ubi non essent ab aliis corporibus gravioribus circumsepta, atque compressa levitatem aliquam demonstrarent. Sed quoniam quaecumque sursum feruntur circumfusa sunt ab aliis corporibus liquidis specie gravioribus, suspicari licet non ab innato principio, sed a violenta ambientis extrusione huiusmodi motum sequi." Ibid., 5r–v.

71. "Democritus etiam, et Plato nostram confirmant sententiam"; "Si aliquis sit, qui nostris rationibus se victum non profiteatur adest Io: Alphonsum Borelli de motibus naturalibus a gravitate factis." Ibid., 5r.

72. "Verum, et quodlibet metallum, auro excepto, in argento vivo demersum attollitur sursum et fluitat, quod sane nemo referat ad innatam levitatem, sed ad extrusionem liquidi omnium gravissimi. In vitreo vase, a quo exhaustus est aer, fumus ope ignis excitatus non sursum tendit, ut in aere, sed praeceps auit deorsum." Ibid., 5v. Criticizing Aristotle's followers but not Aristotle himself was common in the period. Galileo famously made such remarks, as did other university professors. For two examples, see, e.g., Baldini, "Tra due paradigmi?," 207; and Galilei, *EN*, 5:235.

73. "De aeris nos ambientis pondere dubitabant Philosophi, sed illud demonstraverat iampridem Aristoteles, ab Aristotelicis postea in re adeo evidenti desertus, et oppugnatus, sed recentiores hoc pondus ostendunt statera, quum sit exploratum vasa vitrea grandiuscula esse graviora dum aere plena sunt, quam quum aer ab ipsis discessit." BML, Giannetti, "Physica," 5v.

74. "Cartesius celebris sectae conditor, quum iuxta sua principia terram diurna vertigine circa suum axem rotari putaret, cuiusdam subtilis materiae vorticem velocissime volvi circa ipsam statuit. . . . Hoc enim figmentum vel ex eo a nobis est repudiandum, quod motum terrae adstruit fidei, et rationi contrarium, ideoque in hoc diutius non immoramur." Ibid., 6v–7r.

75. "Illustris Gassendus opinatus est, totum terrae globum esse ingentem magnetem, quae fuit Gilberti Angli sententia. . . . Haec opinio ingenii plurimum, sed soliditatis parum continet, ut ostendit praeclare Borellus in libro de vi percussionis." Ibid., 7r–v.

76. Iofrida, "La filosofia e la medicina," 316–22; Bertoloni Meli, *Mechanism, Experiment, Disease,* 160.

77. One section of notes, devoted primarily to medical and anatomical topics, is organized alphabetically with letter tabs cut out from the margin of the pages for easier finding. BML, Bellini, "Opere Varie," 1:164r ff. The second, which is titled "Excerpta philosophica ex variis Auctoribus," focuses on topics related to natural philosophy. Ibid., 4:182r ff. The last, entitled "Studia et Citationes," is more detailed and less organized. Ibid., 6:39r ff.

78. "Aqua. Aquam secundum quid grave asserit Arist. IV. de Coelo pluribus in locis. // Aer. Aer secundum quid levis est per eundem ibidem. / Aer frigidus ex toto est . . . apud Plutarchum in lib. de primo frigido, et primum frigidum . . . dicitur. // Aqua. Aqua per Empedoclem, ac Stratonem est primum frigidum apud eundem. / In aqua continere omnia saporum semina, sed ob . . . insensilia tenet Empedocles apud Gass. v. IV. lib. VI. sect. prima." Ibid., 4:183r.

79. Yeo, "John Locke's 'New Method' of Commonplacing"; Yeo, *Notebooks, English Virtuosi, and Early Modern Science,* 176–82.

80. "Motus et Motor," in BML, Bellini, "Opere Varie," 6:280r.

81. "Motus in naturelem ac violentum dividi potest. Naturalis est qui natura vel sine repugnantia fit. Violentus vero qui praeter natura, vel cum aliqua repugnantia." Ibid., 4:280r. "Non potest . . . [illegible] moveri, nisi et motor ipse moveatur (loquendo de Motoribus secundis ac finitis, non de Primo, et infinito) quia quodcumque movet agit, actio autem motus est et passio item motus, immo actio et passio cum ipso motu identificantur per Arist. 'Phys. t. XVI, et XVII.'" Ibid., 4:281r.

82. "Proinde accelerationem motus fieri secundum proportionem numerorum imparum ab unitati crescentium seu quod idem, secundum ipsarum quadrata adeo proinde ut si prima pulsatione arteriae unum spatium excurritur a mobili in secundo tria excur-

ranter, in tertio quinque, in quarto septem. etc. Quae ostendit Galil. in suo dial. 'de motu Unif. accel'. . . . Linea vero quae a proiecto describitur in descensu Parabolica, est ut Galil. Caval. ac Torricell. ostendant." Ibid., 4:280v–281r.

83. "Causa huius accelerationis referes, ut dictum est in attractione terrae, quod sic clarissime explicatur a Gass. v.III. lib V. sect. prima." Ibid., 4:280v.

84. Maffei, "Relazione di Giovanni Maffei," transcription in Galluzzi, "La scienza davanti alla Chiesa e al Principe," 1325.

85. Ibid., 1325–26.

86. Feingold, "Mathematical Sciences and New Philosophies," 402.

87. Maffei, "Relazione di Giovanni Maffei," transcription in Galluzzi, "La scienza davanti alla Chiesa e al Principe," 1326.

88. Ibid., 1327.

89. Ibid., 1328.

90. "Lettera di G. A. Borelli a Leopoldo dei Medici," transcription in Galluzzi, "La scienza davanti alla Chiesa e al Principe," 1333.

91. Giovanni Maffei taught logic from 1635 to 1638, then served as extraordinary professor of philosophy from 1638 to 1651 and ordinary professor of philosophy from 1651 to 1678. Andrea Moniglia, who served as the doctor to the grand duchess, later taught medicine and "de morbis mulierum" from 1667 to 1700. Luca Terenzi taught logic from 1653 to 1657, then served as a professor of philosophy and practical and theoretical medicine until 1697. Bellini began his teaching at Pisa only in 1663 as a reader in logic, then taught philosophy and anatomy until 1704. Rossetti taught logic from 1666 to 1668 and then served as extraordinary professor of philosophy from 1668 to 1674. Marchetti taught logic from 1659 to 1660, philosophy from 1660 to 1677, and mathematics from 1677 to 1714. See Barsanti, "I docenti e le cattedre dal 1543 al 1737." On Moniglia's appointment as doctor to the grand duchess, see Galluzzi, "La scienza davanti alla Chiesa e al Principe," 1322.

Chapter 6: Jesuit Bookish Practices Applied to the *Two New Sciences*

1. For an overview of recent scholarship on the Jesuits that focuses on their intellectual and cultural activities, see O'Malley et al., *Jesuits*; and O'Malley et al., *Jesuits II*.

2. Sommervogel, *Bibliothèque de la Compagnie de Jésus*, 5:765–69; Zanfredini, "Mauro, Silvestro."

3. Marenbon, *Later Medieval Philosophy*, 7–34.

4. Here I mention the works that contain material relevant to the second and third years of the philosophical sequence. Personal inspection revealed that the numerous commentaries on Aristotle's logic, which made up the first year of the curriculum, are irrelevant for a study of the incorporation of Galileo's *Two New Sciences* in natural philosophical teaching. With respect to the works that address material taught in the second and third years of the curriculum, these manuscripts are not evenly distributed across the seventeenth century but are instead concentrated in the period after 1660. Four manuscript sources date from 1625 to 1629. Only two manuscripts are extant from the period between 1638 and 1660, while two additional manuscripts can be dated precisely to 1660. A third text, while undated, can be assigned to the 1660s on the basis of its attribution to Dominico Marini, who taught philosophy at the Collegio from 1661 to 1666. The remaining five manuscripts date from 1680 to 1696. A final

undated manuscript is attributed to Panici, who taught at the Collegio Romano from 1657 to 1716.

5. For example, the lectures on Aristotle's *Physics* given by the professor of natural philosophy Augustinus de Mari survive in two copies, one of which is a copy made by a student, the other possibly that of de Mari himself. APUG, FC 55, FC 1513.

6. On the development and origins of the *Ratio Studiorum*, see Villoslada, *Storia del Collegio Romano*, 84–115; and Mir, *Aux sources de la pédagogie des Jésuites*.

7. On this trend in the Jesuit curriculum, see Hellyer, *Catholic Physics*, 74–77.

8. Paul Grendler has found that book prices in Venice varied widely depending on size and length, with smaller books tending to cost less than larger, longer ones. Grendler, *Roman Inquisition and the Venetian Press*, 12. The 1703 *Bibliotheca Hookiana*, the auction catalogue of Robert Hooke's library, also indicates that the books that sold for the highest prices were folios, but many quarto books were more expensive than some folios. For an online transcription, see Poole, Henderson, and Nasifoglu, "Hooke's Books Database / Robert Hooke's Books."

9. Grafton and Jardine, *From Humanism to the Humanities*, 111.

10. Lines, "Teaching Physics in Louvain and Bologna," 183, 185.

11. Mauro, *Quaestionum Philosophicarum . . . Libri Tres*, 3:400. The relevant passage in the *Two New Sciences* is Galilei, *EN*, 8:121.

12. Mauro, *Quaestionum Philosophicarum . . . Libri Tres*, 3:402.

13. Ibid., 3:404.

14. Ibid., 3:407.

15. Galilei, *EN*, 8:124.

16. Mauro, *Quaestionum Philosophicarum . . . Libri Tres*, 3:407.

17. Ibid., 3:415–16.

18. Baldini, *Legem Impone Subactis*, esp. 19–119; Baldini, "Development of Jesuit 'Physics' in Italy"; Hellyer, *Catholic Physics*, esp. 13–89.

19. Hellyer, *Catholic Physics*, 90–113, 138–61.

20. A second edition was published in 1670, divided into five volumes. Sommervogel, *Bibliothèque de la Compagnie de Jésus*, 5:765–67. For Mauro's influence on his successors, see the example of Ignatius Tellin discussed below.

21. Mauro, *Quaestionum Philosophicarum . . . Libri Tres*, 3:415–16.

22. Ibid.

23. Galilei, *EN*, 8:210, 212–14.

24. Mauro, *Quaestionum Philosophicarum . . . Libri Tres*, 3:32–33.

25. Ibid., 3:38.

26. Ibid., 3:42. Mauro did not make this distinction between "Galilean" and non-Galilean Copernicans; he referred only to the "Copernican hypothesis." I include the adjective *Galilean* to make clear that this argument from the tides originated not with Copernicus but with Galileo.

27. Because the distance traveled is proportional to the square of the time, the change in velocities between equal intervals of time augment at a constant rate, while the spaces increase according to the odd numbers.

28. Mauro, *Quaestionum Philosophicarum . . . Libri Tres*, 3:455–56.

29. On Baldigiani, see Findlen, "Living in the Shadow of Galileo."

30. "Galileus de facto in suis libris de motu locali demonstrat quod grauia naturaliter et libere intra idem medium descendentia augebunt pro singulis instantibus siue pro singulis temporis partibus aequalibus suam uelocitatem, et motum ita ut uelocitates, seu quod idem est, spatia percursa a mobili crescant pro singulis instantibus proportione aritmetica illa qua numeri impares crescunt in serie naturali nempe 1, 3, 5, 7, 9, 11, hinc bene infert quod uelocitates, et motus siue spatia percursa in tali descensu comparata ad inuicem erunt semper in ratione duplicata temporum, per quae fit motus, siue [quod idem est ut talium temporum quadrata]. ex hoc principio et regula fundamentali deducit plures et mirabiles conclusiones ad motum spectantes." Biblioteca Casanatense, MS 1203, 39v, emphasis added.

31. When he broaches the subject of naturally accelerated motion, the Academician refers to his previous discussion of uniform motion as comprising "the preceding book." Galilei, *EN*, 8:190, 197.

32. "Galileus in dialogis de mundi sistemate." Biblioteca Casanatense, MS 1203, 19r.

33. "Sit perforata terra a superficie usque ad centrum quod iuxta communem calculum distat a nobis per milliaria italica 4000 supponatur autem graue aliquod libere descendens spatio unius minuti secundi temporis descendere per tres passus Geometricos quaeritur quanto tempore graue descendet a superficie terrestri ad centrum descendendo per numeros passuum impares incipiendo ab unitate iuxta regulam datam libere descendentium." Ibid., 47r–v.

34. Galilei, *EN*, 7:245–51.

35. Galilei, *Dialogo sopra i due massimi sistemi del mondo Tolemaico e Copernicano*, 2:532–33.

36. On note-taking in Jesuit teaching, see Nelles, "Libros de Papel, Libri Bianchi, Libri Papyracei." On how this culture of note-taking shaped Jesuits' missionary activities outside the classroom, see Nelles, "Seeing and Writing."

37. Hellyer, *Catholic Physics*, 156, 158–61.

38. On Semery, see Sommervogel, *Bibliothèque de la Compagnie de Jésus*, 7:1115–16.

39. Semery, *Triennium Philosophicum*, 3:395.

40. This debate and its reception in the historiography are described in chapter 1.

41. Galileo stated his definition of naturally accelerated motion twice: Galilei, *EN*, 8:198, 205.

42. This language is found in Proposition 2 and its first corollary in Day 3. Ibid., 8:209–10.

43. Cabeo, *In Quatuor Libros Meteorologicorum Aristotelis Commentaria*, 1:91. For Cabeo's notion of impetus, see Borgato, "Niccolò Cabeo tra teoria ad esperimenti," 365–75.

44. Cabeo, *In Quatuor Libros Meteorologicorum Aristotelis Commentaria*, 3:40.

45. Ibid., 1:94.

46. Ibid., 3:40.

47. Borgato, "Niccolò Cabeo tra teoria ad esperimenti," 370–75.

48. Cattaneo referenced Cabeo's work in his discussion of atomism. See Cattaneo, *Cursus Philosophicus*, 2:25. Mauro also cited Cabeo's commentary in his discussion of heaviness and lightness, as well as in his refutation of Galileo's theory of the tides. Mauro, *Quaestionum Philosophicarum . . . Libri Tres*, 3:42, 426–27.

49. Dinis, "Was Riccioli a Secret Copernican?"; Aricò, "Riccioli nella cultura bolognese del suo tempo."

50. Riccioli, *Almagestum Novum*, 2:383.

51. Cabeo, *In Quatuor Libros Meteorologicorum Aristotelis Commentaria*, 1:88.

52. Throughout his discussion of Galileo's science of motion, Riccioli only referenced the *Dialogue*. Riccioli, *Almagestum Novum*, 1:84–91; 2:313–14, 381–428. Riccioli's citations are to specific pages in the *Dialogue*, almost always indicating whether he was citing the 1632 Italian or the 1635 Latin text. Examination of other sections of his *Almagestum Novum* where we might expect mention of the *Two New Sciences*, for example, in discussions of air, reveals that Riccioli also did not cite the *Two New Sciences*. See chapter 1 for further discussion.

53. Galilei, *EN*, 8:123–24.

54. On why this type of account is problematic for philosophical demonstrations, see Dear, *Discipline and Experience*, 21–30.

55. Mauro, *Quaestionum Philosophicarum . . . Libri Tres*, 3:404.

56. Ibid., 3:406–7, emphasis added.

57. Ibid., 3:407.

58. On the difficulties of integrating experimental evidence in natural philosophical demonstrations, see Dear, *Discipline and Experience*, 32–62.

59. For a similar argument, see Blair's claim that Bodin's use of the commonplace method allowed him to make "a 'matter of fact' out of someone else's evidence." Blair, *Theater of Nature*, 75.

60. Sommervogel, *Bibliothèque de la Compagnie de Jésus*, 7:1920.

61. "7a propositio: de reliquis elementis et mixtis. probabile est, quod non habent positivam levitatem, sed solam gravitatem minorem, quod ex eo patet, quod nunquam corpus ascenderet supra gravius in specie, nisi cum illis imprimitur impetus a graviore extrudente, quod quidem patet ex 4. de Caelo t.39. nam cum uter inflatus, adeoque aere plenus demergitur in aqua, magno impetu conatur ascendere, si illa aqua in qua est demersus esset conglaciata, quia non extruderetur, maneret quietus sub aqua graviore, nec haberet impetum ascendendi, immo potius conatus glaciem deprimeret et rumperet, sicut et aer demersus in concavitibus lapidum et metallorum inclusus exercet impetum ad ascendendum. His accedit, quod utres inflati, ut habetur 4. Coeli t. 30. magis ponderant quam non inflati. ergo signum est aerem simplicem esse gravem et ponderosum, ita ut Galilaeus deprehendit, gravitatem aquae ad gravitatem aeris esse in proportione quadringentorum ad unum." APUG, FC 1344, 162.

62. Like Mauro, Tellin deemed it "probable" that all elements except fire and other mixtures were heavy. Tellin's initial statement regarding the behavior of these elements and mixtures, as well as the four proofs upon which he relied, were also found in Mauro's text.

63. Cattaneo, *Cursus Philosophicus*, 3:234.

64. Tolomei, *Philosophia Mentis*, 531.

65. Ibid.

66. Sommervogel, *Bibliothèque de la Compagnie de Jésus*, 6:166.

67. "Circa gravitatem, et gravitationem aeris duo quaeri possunt. Primo quanta sit eius gravitas . . . Gravitas aeris non potest commodius explicari quam per comparationem

ad gravitatem aliorum corporum, quorum gravitatem per sensum experimur. Si aer igitur comparetur cum aqua in ratione ponderis, se habebit ut unum ad quingenta, si credimus Galilaeo, vel si credimus Mersennio, Fabrio, aliisque, ut unum ad mille: e.g. si palmus aeris ponderet unam libram, palmus aquae ponderabit quingentas, aut mille libras." APUG, FC 1093, 412v.

68. "Supponendum primo cum Aristotele aeri convenire suam propriam gravitatem per quam gravitat in corpora sibi subiecta, quamvis eius gravitationem, seu pressionem non percipiamus: quemadmodum neque natatores in fundo maris sensu percipiunt gravitationem, seu pressionem, quam de facto habet moles aquae superincumbentes. Verba Aristotelis in hac materia apertissima sunt libro 4. de coelo textu 29. *In sua regione omnia gravitatem habent, praeter ignem, etiam aer ipse signum huius est, quia ponderat plus uter inflatus, quam vacuus.* Adverte tamen, in utre inflato contineri aerem vi adactum, et compressum: adeoque posse ponderari in aere externo non adeo compresso. Et per hoc praevertitur aliquorum obiectio, quod aer in aerem ponderari non possit." Ibid., emphasis in original.

69. "Dicunt aliqui esse causam physicam, eamque efficientem, vel exigentem, cui nomen fecerunt Fuga, sive horror Vacui. Volunt proinde esse vel accidens rebus superadditum, ut gravitas innata: vel virtutem identificatam cum corporibus, quae in statutis circumstantiis faciat motum gravium sursum, et levium deorsum, vel etiam in utrisque impediat motum illis connaturalem, si ex tali motu contingeret periculum Vacui. Ista seu virtus, seu qualitas admitti non potest. Primo, quia deberet poni in lapide duplex virtus, alia tendendi sursum, alia tendendi deorsum: ita ut lapis ingenio suo deberet discernere, quando deberet adhibere hanc potius, quam illam virtutem: quod est sane ridiculum. 2.º Quia deberet ista qualitas esse valde callida, et ingeniosa, ut discerneret inter tubos longiores, et breviores: inclinatos et non inclinatos: positos in montibus, vel in vallibus: rursus discerneret Aquilonem ab Austro: unam temperiem aeris ab alia; et pro horum omnium diversitate, diversos motus producere in corporibus." Ibid., 414r.

70. Galilei, *EN*, 8:109.

71. "3.º Quia etiam admissa tali vel virtute, vel qualitate, adhuc plura explicari non possunt; nam si ascensus gravium e.g. haberetur praecise ex horrore Vacui, sequeretur quod corpora possent sursum ascendere semper magis, et magis, neque daretur certa aliqua mensura altitudinis, ultra quam per nullas vires naturales elevari possunt si quidem semper militaret ratio impediendi Vacuum. Hoc autem esse falsum manifeste deprehendit Galilaeus, et constat quotidiana experientia, qua compertum est, aquam in anthliis ascendere solum posse, usque ad altitudinem cubitorum circiter octodecim, ultra quam, nulla vi attractiva, potest sursum elevari. . . . Alia etiam plura similia numerari possent, sed ista sufficiant." APUG, FC 1093, 414r–v. For Galileo's observation, see Galilei, *EN*, 8:64.

72. For a description of the range of these discussions, see Raphael, "Copernicanism in the Classroom."

73. "Prima eius inuentio debetur casui ut narrat Cartesius in sua Dioptrica. Subsequens eius expolitio et perfectio debentur Galileo, qui rude illius inuentu adeo excoluit ut a pluribus habeatur tamquam Author et inuentor Telescopii"; "Horum non sia peruenit etiam ad Galileum, qui rude hoc inuento adeo excoluit et ad tale perfectione perduxit." APUG, FC 269, 39r, 40r.

74. "Depraehenderunt insuper circa Iovem quatuor stellas errare motibus diversis, quas Galileus, qui eas primus observavit sui principis stemmati alludens stellas Mediceas nominavit." APUG, FC 1347, 326r.

75. Mauro, *Quaestionum Philosophicarum . . . Libri Tres*, 3:32–33.

76. See, e.g., Cattaneo, *Cursus Philosophicus*, 2:753–54. "R. contra Nicolaum Copernicum. Galilaeo de Galilaeis, quos sequuntur non nulli recentiores, quod in medio mundo non est sol immotus, circa quem movet moles terraquea, sed circa hanc immobilem et universam concentricam movet sol." APUG, FC 1513, 306v.

77. Eschinardi, *Cursus Physicomathematicus*, 175.

78. Raphael, "Copernicanism in the Classroom."

79. For some key examples, in addition to those listed below, see Feingold, *Mathematicians' Apprenticeship*; and Porter, "Scientific Revolution and Universities."

80. For key examples of this literature, see Siebert, "Kircher and His Critics"; and Stolzenberg, "Oedipus Censored." A similar impulse underlies scholarship that seeks to identify the "true" affinities of individual Jesuits and that examines the Jesuits' formal and informal censoring practices. See, e.g., Baldini, *Legem Impone Subactis*, 75–119, 251–81; Dinis, "Was Riccioli a Secret Copernican?"; Hellyer, "Because the Authority of My Superiors Commands"; and Hellyer, *Catholic Physics*, 35–52.

81. This argument was articulated with force in the late nineteenth century, when the Jesuit order was thought not only to have been a nonparticipant in the transformations of early modern science but even to have actively thwarted them. It still finds currency today in scholarship that differentiates between "conservative" and "innovative" Jesuits on the basis of whether individuals rejected (or embraced or tried to embrace) novelty and whether they restricted (or encouraged) its spread within the order's scholarship. For example, in his analysis of late seventeenth-century Jesuit natural philosophers in what is now Germany, Marcus Hellyer details the sparse mentions of work on air pumps in Jesuit teaching texts, as well as professors' rejection of the claims of those who carried out such experiments and their overwhelming adherence to a strictly traditional Aristotelian-Peripatetic explanation. He concludes that Jesuit professors' earlier "willingness to engage with novelties . . . had faded." Hellyer, *Catholic Physics*, 161. Peter Dear stresses the contrast between Jesuit mathematicians, who aimed to promote their discipline and its approaches, and their "adversaries," the order's natural philosophers, who "denied to mathematics the status of *scientiae*, true scientific knowledge." Dear, *Discipline and Experience*, 36. John Heilbron described the opposition from conservative Jesuits in the following terms: "Opposition from colleagues posed another obstacle. Simplicios did exist, and exercised their influence both unofficially, through local cabals, and institutionally, through the censorship of the press. The irregular practices were probably the more effective: conservative rectors or prefects of studies might burden their modernizing professors with routine assignments or secure their transfer to non-academic positions." Heilbron, *Electricity in the 17th and 18th Centuries*, 113.

82. Similar arguments can be found in recent discussions of Jesuit contributions to science. On the need to recognize the contributions of the Catholic Church to science in the period, Heilbron writes, "The work of the meridian makers shows that men whose careers were underwritten in whole or part by the Church could contribute importantly to the development of astronomy, that is, to the leading sector of natural

knowledge during the seventeenth century." Heilbron, *Sun in the Church*, 20. One piece of evidence that justified his description of Riccioli as "the predominant authority among the strong-minded mathematicians in Bologna" was the fact that he "belonged to the first generation of Jesuits [who] taught astronomy in a moderately modern manner." Ibid., 86. Peter Dear gives the "enormous role of Jesuit scholarship in the mixed mathematical sciences throughout this period" as his reason for assigning Jesuit mathematicians a central role in his account of the rise of physicomathematics. Dear, *Discipline and Experience*, 6–7. Christoph Clavius deserves our attention because, among other reasons, his "astronomical work is significant in its own right"; he "adapted some of Copernicus's innovations . . . to the Ptolemaic framework"; he was "one of the earliest and most influential endorsers of Galileo's telescopic discoveries—though he stopped far short of agreeing with Galileo's interpretations"; and his influence was "nowhere greater than in his role as a textbook author." Lattis, *Between Copernicus and Galileo*, 4. Note that many of these same historians at times make the opposite argument, claiming that other members of the Jesuit order were "conservative." See the previous note.

83. For a fuller discussion of these issues in a very different context, see Kors, *Orthodox Sources of Disbelief*, 81–109.

84. Nelles, "Libros de Papel, Libri Bianchi, Libri Papyracei," 101.

85. On these and other questions in the medieval tradition, see Grant, *Much Ado about Nothing*.

86. On the standardization of topics set by Aristotle's text and his commentators in the period, see Baldini, "Development of Jesuit 'Physics' in Italy," 252; Brockliss, "Curricula," 580; Brockliss, *French Higher Education in the Seventeenth and Eighteenth Centuries*, 337; and Grendler, *Universities of the Italian Renaissance*, 277.

87. Hellyer, *Catholic Physics*, 138–61.

88. The two exceptions are Panici and Baldigiani, who cited the measurement in support of those who argued that the heaviness of the air explained recent experiments with inverted tubes of mercury and the air pump.

89. As one typical example, consider the teaching text of Fabio Ambrogio Spinola (1593–1671) for a course he gave in 1626 that was recorded by Gulielmo Roncallo Pomaranceo. Spinola divided his *quaestio* on the elements into six doubts (*dubitationes*), which he titled, "An et quot sint elementa," "Quot et quid sint primae qualitates actiuae," "Quomodo primae qualitates conveniant singulis elementis," "Quot et quae sint uirtutes motiuae in elementis," "A quo moueantur grauia et leuia . . . et cur uelocius in fine," and "An elementa transmutari possint ad inuicem." APUG, FC 520, 178v, 182r, 186v, 191v, 194r, 200v.

90. On Galileo's familiarity with this tradition, see Camerota and Helbing, "Galileo and Pisan Aristotelianism"; Carugo and Crombie, "Jesuits and Galileo's Ideas of Science and of Nature"; Crombie and Carugo, "Sorting Out the Sources"; and Wallace, *Galileo and His Sources*. On Galileo's attention to this tradition in the composition of his *Two New Sciences*, see Raphael, "Making Sense of Day 1 of the *Two New Sciences*."

91. Galilei, *EN*, 8:121. The example of the inflated pouches cited by Galileo is found in book 4 of *De caelo*, 311b.10–11.

92. Galle, *Peter of Auvergne*, 113, 257–60. For Simplicius's discussion, see Simplicius, *Commentaria in Quatuor Libris Aristotelis De Coelo*, 710–11.

93. Buridan, *Quaestiones Super Libris Quattuor de Caelo et Mundo*, 265, 268.

94. Ibid., 268.

95. Albert of Saxony, *Quaestiones in Aristotelis De Caelo*, 464, 469.

96. *Commentarii Collegii Conimbricensis Societatis Iesu*, 542.

97. Ibid., 542–43.

98. On such discussions at Pisa, see Camerota and Helbing, "Galileo and Pisan Aristotelianism."

Epilogue

1. See, e.g., Galileo's displeasure at Cavalieri's publishing his findings on the parabolic trajectory of projectiles before he himself did. Drake, *Galileo at Work*, 340.

2. Galluzzi, "Sepulchers of Galileo."

3. Flitner, *Erasmus im Urteil seiner Nachwelt*, 110–29; Mansfield, *Phoenix of His Age*, 231–58; Blayney, *First Folio of Shakespeare*. On Viviani's awareness of these trends, see Wilding, "Return of Thomas Salusbury's *Life of Galileo* (1664)," 253.

4. On the construction of the Galileo archive and Viviani's motivations, see Bucciantini, "Celebration and Conservation." For examples of Viviani's efforts to monitor contemporaries' writings about Galileo, see Findlen, "Living in the Shadow of Galileo," 244–51; and Wilding, "Return of Thomas Salusbury's *Life of Galileo* (1664)," 252–54.

5. On Viviani's report and the scholarship devoted to assessing its reliability, see Camerota and Helbing, "Galileo and Pisan Aristotelianism," 319–21.

6. Bucciantini, "Celebration and Conservation." On Favaro, see Bucciantini, "Favaro, Antonio," and sources cited therein.

7. Favaro, *Amici e corrispondenti di Galileo*; Favaro, "Gli oppositori di Galileo: I. Antonio Rocco"; Favaro, "Gli oppositori di Galileo. II. Liberto Froidmont"; Favaro, "Gli oppositori di Galileo. III. Cristoforo Scheiner." On Favaro's scholarship, see Favaro, *Amici e corrispondenti di Galileo*, 1:v–xi.

8. Koyré, "Experiment in Measurement."

9. Koyré, "Documentary History of the Problem of Fall."

10. Drake, *History of Free Fall*, 67–79.

11. See, e.g., Galluzzi, "Gassendi and l'Affaire Galilée of the Laws of Motion." Subsequent studies have returned to this narrative and set of historical actors. See Elazar, *Honoré Fabri and the Concept of Impetus*; Elazar and Feldhay, "Honore Fabri S.J. and Galileo's Law of Fall"; Garber, "On the Frontlines of the Scientific Revolution"; Palmerino, "Galileo's and Gassendi's Solutions"; Palmerino, "Galileo's Theories of Free Fall and Projectile Motion"; Palmerino, "Gassendi's Reinterpretation of the Galilean Theory of Tides"; Palmerino, "Infinite Degrees of Speed"; and Palmerino, "Two Jesuit Responses to Galileo's Science of Motion."

12. One notable exception has been the work of Carla Rita Palmerino. See, e.g., Palmerino and Thijssen, *Reception of the Galilean Science of Motion*; Palmerino, "Two Jesuit Responses to Galileo's Science of Motion"; Palmerino, "Infinite Degrees of Speed"; and Palmerino, "Gassendi's Reinterpretation of the Galilean Theory of Tides."

13. Two of the most well known examples of studies of early modern science that have focused on conflict are Biagioli, *Galileo, Courtier*; and Shapin and Schaffer, *Leviathan and the Air-Pump*. Consider also the more recent Dascal and Boantza, *Controversies within the Scientific Revolution*.

14. An example is Baldini, "Tra due paradigmi?"

15. For some key examples, see Brockliss, *French Higher Education in the Seventeenth and Eighteenth Centuries*, 341–42; Dinis, "Was Riccioli a Secret Copernican?"; and Hellyer, *Catholic Physics*, 6–7.

16. Clucas, "Galileo, Bruno, and the Rhetoric of Dialogue," 411–12; Raphael, "Galileo's *Two New Sciences* as a Model of Reading Practice."

17. In the words of Elizabeth Yale, "A special affinity grew up between Baconian 'facts' and manuscript—often in the form of commonplace books and loose leaf notes—as the material means of collecting, recording, organizing, and sharing them within these fields." Yale, "Marginalia, Commonplaces, and Correspondence," 195. On Baconianism and textual practices, see Yeo, "Between Memory and Paperbooks."

18. Grafton, *Defenders of the Text*, 5.

Primary Sources

Manuscripts

Florence, Biblioteca Medicea Laurenziana (BML)

Bellini, Lorenzo. "Opere Varie." MS Ashburnham (Ashb.) 638.

Giannetti, Pascasio. "Physica." MS Ashb. 872.

Mancini, Mauro. "De Physica." MS Ashb. 1623.

Florence, Biblioteca Nazionale Centrale di Firenze (BNCF)

MS Galileo (Gal.) 74. "Opere di Galileo Galilei, Parte 5ª, Tomo 4, Meccanica."

MS Gal. 114. Renieri, Vincenzio. "Vol. 2. Matematica pura applicata."

MS Gal. 115. Renieri, Vincenzio. "Vol. 3. Astronomia e Carteggio scientifico."

Paris, Bibliothèque nationale de France (BnF)

MS Français (Fr.) 12357. Anon. "Remarques tirées du livre de l'Harmonie universelle du P. MERSENNE, ainsy qu'il les avoit escrites, de sa main, à la marge et aux feuillets blancs devant et derrière dudit livre."

Rome, Archivio Storico della Pontificia Università Gregoriana (APUG)

Fondo Curia (FC) 55. de Mari, A. "Physica."

FC 269. Febei, Francesco Antonio. "Tractatus Geometriae Practicae."

FC 520. Spinola, F. A. "In Libros Aristotelis de Generatione et Corruptione Auditore Gulielmo Roncallo Pom.cio."

FC 1093. Panici, J. J. "In Libros Aristotelis de Physico Auditu."

FC 1344. Tellin, I. "Tertia Pars Philosophiae Peripateticae."

FC 1347. Bompiani, Ludovico. "Disputationes Physicae."

FC 1513. de Mari, A. "Physica P. Augustini de Mari Excepta à I.B.S.I."

Rome, Biblioteca Casanatense

MS 1203. Baldigiani, A. "Exercitationes Physicomathematicae et Progressiones Mathematicae."

Annotated Printed Books

Florence, Biblioteca Nazionale Centrale di Firenze (BNCF)

MS Banco Rari (B. Rari) 169. Galilei, Galileo. *Discorsi e dimostrazioni matematiche intorno à due nuoue scienze, attenenti alla mecanica & i movimenti locali.* Leiden, 1638.

MS Gal. 79. Galilei, Galileo. *Discorsi e dimostrazioni matematiche intorno à due nuoue scienze, attenenti alla mecanica & i movimenti locali.* Leiden, 1638.

MS Gal. 80. Galilei, Galileo. *Discorsi e dimostrazioni matematiche intorno à due nuoue scienze, attenenti alla mecanica & i movimenti locali.* Leiden, 1638.

Oxford, Bodleian Library, University of Oxford

Savile A.19. Galilei, Galileo. *Opere di Galileo Galilei Linceo nobile Fiorentino.* 2 vols. Bologna, 1655–56.

Savile Aa.12. Davison, William. *Philosophia Pyrotechnica, seu Curriculus Chymiatricus.* 4 parts in 3 vols. Paris, 1635.

Savile Bb.13. Galilei, Galileo. *Discorsi e dimostrazioni matematiche intorno à due nuoue scienze, attenenti alla mecanica & i movimenti locali.* Leiden, 1638.

Savile X.9(2). Apollonius of Perga, Federigo Commandino, Pappus of Alexandria, and Antinoensis Serenus. *Conicorum Libri Quattuor. Unà cum Pappi Alexandrini Lemmatibus, et Commentariis Eutocii Ascalonitae. Sereni Antinsensis . . . Libro Duo. Quae omnia nuper F. Commandinus e Gr. Convertit. et Comm. Illustravit.* Bologna, 1566.

Paris, Bibliothèque de l'Institut de France (BIF)

4° M 541. Galilei, Galileo. *Discorsi e dimostrazioni matematiche intorno à due nuoue scienze, attenenti alla mecanica & i movimenti locali.* Leiden, 1638.

Paris, Bibliothèque Sainte-Geneviève (BSG)

4V 585 INV 1338 RES. Galilei, Galileo. *Discorsi e dimostrazioni matematiche intorno à due nuoue scienze, attenenti alla mecanica & i movimenti locali.* Leiden, 1638.

Printed Primary Sources

Albert of Saxony. *Quaestiones in Aristotelis De Caelo.* Edited by Benoît Patar. Louvain-la-Neuve, Belgium: Editions de l'Institut supérieur de philosophie, 2008.

Baliani, G. B. *De Motu Naturali Gravium Solidorum.* Genoa: Typographia Io. Mariae Farroni, Nicolai Pesagnij & Petri Francisci Barberij, 1638.

———. *De Motu Naturali Gravium Solidorum et Liquidorum.* Genoa: Typographia Io. Mariae Farroni, 1646.

———. *De Motu Naturali Gravium Solidorum et Liquidorum.* Edited by G. Baroncelli. Florence: Giunti, 1998.

Bérigard, C. *Circulus Pisanus.* 4 parts in 1 vol. Pt. 1, *In Priores Libros Phys. Arist*; pt. 2, *In Octauum Librum Physicorum Aristotelis*; pt. 3, *In Aristotelis Lib. de Ortu & Interitu.* Udine: Nicolae Schiratti, 1643.

———. *Circulus Pisanus . . . De Veteri et Peripatetica Philosophia in Aristotelis Libros de Coelo.* Udine: Nicolai Schiratti, 1647.

———. *Circulus Pisanus.* Pt. 1, *De Veteri et Peripatetica Philosophia in Aristotelis Libros Octo Physicorum.* Padua: Typis Frambotti Bibliopolae, 1661.

Bibliotheca Hookiana: Sive Catalogus Diversorum Librorum. London: E. Millington, 1703.

Birch, Thomas. *The History of the Royal Society of London.* 4 vols. London: printed for A. Millar, 1756–57.

Buridan, Jean. *Quaestiones Super Libris Quattuor de Caelo et Mundo.* Edited by Ernest Addison. New York: Kraus Reprint, 1970.

Cabeo, N. *In Quatuor Libros Meteorologicorum Aristotelis Commentaria.* 4 vols. Rome: Typis haeredum Francisci Corbelletti, 1646.

Catalogus Librorum Diversis Italiae Locis Emptorum Anno Dom. 1647: A Georgio Thomasono Bibliopola Londinensi apud Quem in Caemiterio D. Pauli Ad Insigne Rosae Coronatae Prostant Venales. London: Typis Johannis Legatt, 1647.

Cattaneo, O. *Cursus Philosophicus.* 4 vols. Rome: Nicolai Angeli Tinassij, 1677.

Cavalieri, Bonaventura. *Lo specchio ustorio, overo, Trattato delle settioni coniche : Et alcuni loro mirabili effeti intorno al lume, caldo, freddo, suono, e moto ancora.* Bologna: Presso Clemente Ferroni, 1632.

Commentarii Collegii Conimbricensis Societatis Iesu, in Quatuor Libros de Caelo, Meteorologicos et Parva Naturalia Aristotelis Stagiritae. Cologne, 1603.

The Compleat Gunner in three parts: Part I. Shewing the Art of Founding and Casting . . . The Composition and Matters of Gunpowders . . . : Part II. Discovers the Necessary Instruments . . . To the Compleating of a Gunner . . . Part III. Shews the Nature of Fire-Works. London: printed for Rob. Pawlet, Tho. Passinger, and Benj. Hurlock, 1672.

Descartes, René. *Discourse on Method, Optics, Geometry, and Meteorology.* Translated by Paul J. Olscamp. Indianapolis: Hackett, 2001.

———. *Oeuvres de Descartes.* Edited by Charles Adam and Paul Tannery. 11 vols. Paris: Librairie Philosophique J. Vrin, 1996.

———. *The Philosophical Writings of Descartes.* Translated by John Cottingham, Robert Stoothoff, and Dugald Murdoch. 3 vols. Cambridge: Cambridge University Press, 1984–91.

Eschinardi, F. *Cursus Physicomathematicus.* Rome: Ex Typographia Ioannis Iacobi Komarek Bohëmi apud Angelum Custodem, 1689.

Galilei, Galileo. *Dialogo sopra i due massimi sistemi del mondo Tolemaico e Copernicano.* Edited by Ottavio Besomi and Mario Helbing. 2 vols. Padua: Editrice Antenore, 1998.

———. *Discorsi e dimostrazioni matematiche.* Edited by Enrico Giusti. Turin: Giulio Einaudi, 1990.

———. *Discorsi e dimostrazioni matematiche intorno a due nuove scienze.* Edited by Adriano Carugo and Ludovico Geymonat. Turin: Paolo Boringhieri, 1958.

———. *Opere di Galileo Galilei.* 2 vols. Bologna: HH del Dozza, 1655–56.

———. *Le opere di Galileo Galilei.* Edited by Antonio Favaro. Edizione nazionale. 20 vols. Florence: G. Barbèra, 1890–1909.

———. *Sidereus Nuncius or The Sidereal Messenger.* Translated by Albert van Helden. Chicago: University of Chicago Press, 1989.

———. *Systema Cosmicum: In Quo Dialogis IV De Duobus Maximis Mundi Systematibus Ptolemaico et Copernicano . . . Ejusdem Tractatus de Motu, Nunc Primum ex Italico Sermone in Latinum Versus.* Leiden: Apud Fredericum Haaring et Davidem Severinum, bibliopolas, 1699.

———. *Two New Sciences Including Centers of Gravity and Force of Percussion.* Translated by Stillman Drake. Toronto: Wall & Thompson, 1989.

Gassendi, P. *Opera Omnia in Sex Tomos Divisa.* 6 vols. Lyon: Sumptibus Laurentii Anisson & Ioan. Bapt. Devenet, 1658. Reprinted with an introduction by Tullio Gregory. Stuttgart–Bad Cannstatt: F. Frommann, 1964.

Grandi, Guido, and Vincenzo Viviani. *Trattato delle resistenze principiato da Vincenzio Viviani per illustrare l'opere del Galileo.* Florence: per G. G. Tartini e S. Franchi, 1718.

Holstenius, Lucas, ed. *Index Bibliothecae qua Franciscus Barberinus, S.R.E. Cardinalis Vicecancellarius Magnificentissimas, suae familiae ad Quirinalem aedes magnificentiores reddidit.* 2 vols. Rome: Typis Barberiis, Excudebat Michael Hercules, 1681.

Hyde, Thomas, ed. *Catalogus Impressorum Librorum Bibliothecae Bodleianae in Academia Oxoniensi.* Oxford: e Theatro Sheldoniano, 1674.

Kircher, Athanasius. *Musurgia Universalis sive, Ars Magna Consoni et Dissoni.* 2 vols. Rome: Ex typographia haeredum Francisci Corbelletti, 1650.

Lipen, Martin. *Bibliotheca Realis Philosophica: Omnium Materiarum, Rerum et Titulorum in Universo Totius Philosophiae. . . .* 2 tomes in 1 vol. Frankfurt: Cura & sumptibus Johannis Friderici, 1682.

Mauro, S. *Quaestionum Philosophicarum Sylvestri Mauri Soc. Iesu. in Collegio Romano Philosophiæ Professoris: Libri Tres Pro Laurea Philosophica Andreae Portner Collegij Germanici, & Hungarici Alumni.* 3 vols. Rome: Ignazio Lazeri, 1658.

Mersenne, Marin. *Cogitata Physico-Mathematica.* Paris: A. Bertier, 1644.

——. *Correspondance du P. Marin Mersenne, Religieux Minime.* Edited by P. Tannery and C. de Waard. 18 vols. Paris: Presses Universitaires de France, 1932–88.

——. *Harmonie universelle contenant la théorie et la pratique de la musique (Paris 1636): Édition facsimilé de l'exemplaire conservé à la Bibliothèque des Arts et Métiers et annoté par l'auteur.* Edited by F. Lesure. 3 vols. Paris: Centre National de la Recherche Scientifique, 1963.

——. *Les méchaniques de Galilée.* Edited by B. Rochot. Paris: Presses Universitaires de France, 1966.

——. *Les nouvelles pensées de Galilée.* Edited by P. Costabel and M. P. Lerner. 2 vols. Paris: Librairie Philosophique J. Vrin, 1973.

——. *Novarum Observationum Tomus III Physico-Mathematicarum.* Paris: A. Bertier, 1647.

——. "Traité des mouvemens et de la cheute des corps pesans & de la proportion de leurs differentes vitesses." *Corpus: Revue de philosophie* 2, no. 2 (1986): 25–58. Originally published in Paris in 1634.

Riccioli, G. B. *Almagestum Novum.* 2 parts in 1 vol. Bologna: Victor Benatus, 1651.

Salusbury, T. *Mathematical Collections and Translations.* 2 vols. London, 1661–65.

Semery, A. *Triennium Philosophicum, Quod P. Andreas Semery Remus, è Societate Jesu, In Collegio Romano Philosophiae Iterum Professor Dictabat. Tertia Hac Editione Quae Est Prima in Germania Correctum.* 3 vols. Cologne: Joannis Caspari Bencardi, 1688.

Simplicius. *Commentaria in Quatuor Libris Aristotelis De Coelo.* Venice, 1584.

Tolomei, J. B. *Philosophia Mentis, et Sensuum Secundum Utramque Aristotelis Methodum Pertractata Metaphysicè, et Empiricè.* Rome: Reverenda Camera Apostolica, 1696.

Viviani, V. *Qvinto libro degli Elementi d'Evclide, ovvero Scienza vniversale delle proporzioni spiegata colla dottrina del Galileo.* Florence: alla Gondotta, 1674.

Ward, Seth, and John Wilkins. *Vindiciae Academiarum, Containing Some Briefe Animadversions upon Mr. Websters Book, Stiled The Examination of Academies.* Oxford: printed by Leonard Lichfield, printer to the University, for Thomas Robinson, 1654.

Secondary Sources

Aricò, D. "Riccioli nella cultura bolognese del suo tempo." In Borgato, *Giambattista Riccioli e il merito scientifico dei Gesuiti*, 251–76.

Ariew, Roger. *Descartes and the Last Scholastics.* Ithaca, NY: Cornell University Press, 1999.

———. "Galileo in Paris." *Perspectives on Science* 12, no. 2 (2004): 131–34.

Baldini, Ugo. "The Development of Jesuit 'Physics' in Italy, 1550–1700: A Structural Approach." In *Philosophy in the Sixteenth and Seventeenth Centuries: Conversations with Aristotle,* edited by Constance Blackwell and Sachiko Kusukawa, 248–79. Aldershot, Hampshire: Ashgate, 1999.

———. *Legem Impone Subactis: Studi su filosofia e scienza dei Gesuiti in Italia, 1540–1632.* Rome: Bulzoni, 1992.

———. "Tra due paradigmi? La 'Naturalis philosophia' di Carlo Rinaldini." In Pepe, *Galileo e la scuola galileiana*, 189–222.

Balsamo, Luigi. *Bibliography: History of a Tradition.* Translated by William A. Pettas. Berkeley, CA: Bernard M. Rosenthal, 1990.

Baroncini, G. *Forme di esperienza e rivoluzione scientifica.* Florence: Olschki, 1992.

Barsanti, Danilo. "I docenti e le cattedre dal 1543 al 1737." In *Storia dell'Università di Pisa,* edited by E. Amatori, vol. 1, pt. 2: 505–67. Pisa: Pacini, 1993.

Beaulieu, A. "Les réactions des savants français au début du XVIIe siècle devant l'héliocentrisme de Galilée." In Galluzzi, *Novità celesti e crisi del sapere*, 373–81.

"Beauregard, Claudio Guillermet." In *Dizionario biografico degli Italiani,* vol. 7, edited by A. M. Ghisalberti, 386–89. Rome: Istituto della Enciclopedia Italiana, 1965.

Bellucci, M. "La filosofia naturale di Claudio Berigardo." *Rivista Critica di Storia della Filosofia* 26 (1971): 363–411.

Bennett, J. A. "Hooke and Wren and the System of the World: Some Points towards an Historical Account." *British Journal for the History of Science* 8, no. 1 (1975): 32–61.

———. *The Mathematical Science of Christopher Wren.* Cambridge: Cambridge University Press, 1982.

Benton, John F. "Appendix: Descartes' *Olympica.*" *Philosophy and Literature* 4, no. 2 (1980): 162–66.

Benvenuto, Edoardo. *An Introduction to the History of Structural Mechanics.* 2 vols. New York: Springer-Verlag, 1991.

Beretta, Marco, Antonio Clericuzio, and Lawrence M. Principe, eds. *The Accademia del Cimento and its European Context.* Sagamore Beach, MA: Science History, 2009.

Bertoloni Meli, Domenico. "Mechanics." In Park and Daston, *Early Modern Science*, 632–72.

———. *Mechanism, Experiment, Disease: Marcello Malpighi and Seventeenth-Century Anatomy.* Baltimore: Johns Hopkins University Press, 2011.

——. *Thinking with Objects: The Transformation of Mechanics in the Seventeenth Century.* Baltimore: Johns Hopkins University Press, 2006.

Biagioli, Mario. *Galileo, Courtier: The Practice of Science in the Culture of Absolutism.* Chicago: University of Chicago Press, 1993.

——. *Galileo's Instruments of Credit: Telescopes, Images, Secrecy.* Chicago: University of Chicago Press, 2006.

Biener, Zvi. "Galileo's First New Science: The Science of Matter." *Perspectives on Science* 12, no. 3 (2004): 262–87.

Blair, Ann. "Humanist Methods in Natural Philosophy: The Commonplace Book." *Journal of the History of Ideas* 53, no. 4 (1992): 541–51.

——. "Note Taking as an Art of Transmission." *Critical Inquiry* 31, no. 1 (2004): 85–107.

——. "Reading Strategies for Coping with Information Overload, ca. 1550–1700." *Journal of the History of Ideas* 64, no. 1 (2003): 11–28.

——. "The Rise of Note-Taking in Early Modern Europe." *Intellectual History Review* 20, no. 3 (2010): 303–16.

——. *The Theater of Nature: Jean Bodin and Renaissance Science.* Princeton, NJ: Princeton University Press, 1997.

——. *Too Much to Know.* New Haven, CT: Yale University Press, 2010.

Blay, Michel. *Reasoning with the Infinite: From the Closed World to the Mathematical Universe.* Translated by M. B. DeBevoise. Chicago: University of Chicago Press, 1998.

Blayney, Peter W. M. *The First Folio of Shakespeare.* Washington, DC: Folger Library, 1991.

Borgato, Maria Teresa, ed. *Giambattista Riccioli e il merito scientifico dei Gesuiti nell'età Barocca.* Florence: Olschki, 2002.

——. "Niccolò Cabeo tra teoria ad esperimenti: Le leggi del moto." In *Gesuiti e università in Europa,* edited by Gian Paolo Brizzi and Roberto Greci, 361–85. Bologna: Cooperativa Libraria Universitaria Editrice Bologna, 2002.

——. "Riccioli e la caduta dei gravi." In Borgato, *Giambattista Riccioli e il merito scientifico dei Gesuiti,* 79–118.

Bougy, Alfred de. *Histoire de la Bibliothèque Sainte-Geneviève.* Paris: Comptoir des Imprimeurs-Unis, 1847.

Brockliss, Laurence. "Curricula." In *Universities in Early Modern Europe (1500–1800),* edited by H. de Ridder-Symoens, vol. 2 of *A History of the University in Europe,* 563–620. Cambridge: Cambridge University Press, 1996.

——. *French Higher Education in the Seventeenth and Eighteenth Centuries: A Cultural History.* Oxford: Clarendon, 1987.

Bucciantini, Massimo. "Celebration and Conservation: The Galilean Collection at the National Library of Florence." In *Archives of the Scientific Revolution: The Formation and Exchange of Ideas in Seventeenth-Century Europe,* edited by Michael Hunter, 21–34. Woodbridge, Suffolk: Boydell, 1998.

——. "Favaro, Antonio." In *Dizionario biografico degli Italiani,* vol. 45, edited by Fiorella Bartoccini and Mario Caravale, 441–45. Rome: Istituto della Enciclopedia Italiana, 1995.

Bucciantini, Massimo, Michele Camerota, and Franco Giudice. *Il telescopio di Galileo: Una storia europea.* Turin: Giulio Einaudi, 2012.

Buccolini, Claudio. "Opere di Galileo Galilei provenienti dalla biblioteca di Marin Mersenne." *Nouvelles de la République des Lettres*, 1998, 139–42.

Butterfield, Herbert. *The Origins of Modern Science*. New York: Free Press, 1965.

Büttner, Jochen, Peter Damerow, and Jürgen Renn. "Traces of an Invisible Giant: Shared Knowledge in Galileo's Unpublished Treatises." In *Largo campo di filosofare*, edited by José Montesinos and Carlos Solís, 183–201. La Orotava, Tenerife: Fundación Canaria Orotava de Historia de la Ciencia, 2001.

Calis, Richard. "Personal Philology." *JHIBlog*. 6 April 2015. http://jhiblog.org/2015/04/06/personal-philology/.

Camerota, Michele. "La biblioteca di Galileo: Alcune integrazioni e aggiunte desunte dal carteggio." In *Biblioteche filosofiche private in eta moderna e contemporanea: Atti del convegno Cagliari, 21–23 aprile 2009*, edited by Francesca Maria Crasta, 81–95. Florence: Casa Editrice Le Lettere, 2010.

———. *Galileo Galilei e la cultura scientifica nell'età della Controriforma*. Rome: Salerno, 2004.

Camerota, Michele, and Mario Helbing. "Galileo and Pisan Aristotelianism: Galileo's 'De Motu Antiquiora' and 'Quaestiones de Motu Elementorum' of the Pisan Professors." *Early Science and Medicine* 5, no. 4 (2000): 319–65.

Carugo, Adriano, and Alistair Cameron Crombie. "The Jesuits and Galileo's Ideas of Science and of Nature." *Annali dell'Istituto e Museo di Storia della Scienza di Firenze* 8 (1983): 3–67.

Clark, Frederic. "Dividing Time: The Making of Historical Periodization in Early Modern Europe." PhD diss., Princeton University, 2014.

Clavelin, M. *The Natural Philosophy of Galileo: Essay on the Origins and Formation of Classical Mechanics*. Cambridge, MA: MIT Press, 1968.

Clucas, Stephen. "Galileo, Bruno, and the Rhetoric of Dialogue in Seventeenth-Century Natural Philosophy." *History of Science* 66 (2008): 405–29.

Cochrane, E. W. "Science and Humanism in the Italian Renaissance." *American Historical Review* 81, no. 5 (December 1976): 1039–57.

Costabel, P., and M. P. Lerner. Introduction to *Les nouvelles pensées de Galilée*, by Marin Mersenne, edited by Costabel and Lerner, 1:15–52. 2 vols. Paris: Librairie Philosophique J. Vrin, 1973.

Costello, William T. *The Scholastic Curriculum at Early Seventeenth-Century Cambridge*. Cambridge, MA: Harvard University Press, 1958.

Craster, H. H. E. *History of the Bodleian Library, 1845–1945*. Oxford: Clarendon, 1952.

Crombie, Alistair Cameron. *Augustine to Galileo*. Cambridge, MA: Harvard University Press, 1961.

———. "Sources of Galileo's Early Natural Philosophy." In *Reason, Experiment and Mysticism in the Scientific Revolution*, edited by M. L. Righini Bonelli and W. Shea, 157–75. New York: Science History, 1975.

Crombie, Alistair Cameron, and Adriano Carugo. "Sorting Out the Sources." *Times Literary Supplement*, 22 November 1985, 1319–20.

Cunningham, Andrew, and Sachiko Kusukawa, eds. and trans. *Natural Philosophy Epitomised: Books 8–11 of Gregor Reisch's* Philosophical Pearl *(1503)*. Farnham, Surrey: Ashgate, 2010.

Damerow, Peter, Gideon Freudenthal, Peter McLaughlin, and Jürgen Renn. *Exploring the Limits of Preclassical Mechanics*. New York: Springer, 2004.

Darnton, Robert. "First Steps toward a History of Reading." In *The Kiss of Lamourette: Reflections in Cultural History*, 154–87. New York: Norton, 1990.

Dascal, Marcelo, and Victor D. Boantza, eds. *Controversies within the Scientific Revolution*. Amsterdam: John Benjamins, 2011.

Daston, Lorraine. "Baconian Facts, Academic Civility, and the Prehistory of Objectivity." In *Rethinking Objectivity*, edited by A. Megill, 37–63. Durham, NC: Duke University Press, 1997.

———. "Taking Note(s)." *Isis* 95 (2004): 443–48.

Davies, D. W. *The World of the Elseviers, 1580–1712*. Westport, CT: Greenwood, 1971.

Dear, Peter. *Discipline and Experience: The Mathematical Way in the Scientific Revolution*. Chicago: University of Chicago Press, 1985.

———. *Mersenne and the Learning of the Schools*. Ithaca, NY: Cornell University Press, 1988.

de Ceglia, Francesco Paolo. "'Additio Illa Non Videtur Edenda': Giuseppe Biancani, Reader of Galileo in an Unedited Censored Text." In Feingold, *New Science and Jesuit Science*, 159–86.

Des Chene, Dennis. *Physiologia: Natural Philosophy in Late Aristotelian and Cartesian Thought*. Ithaca, NY: Cornell University Press, 2000.

Dinis, A. "Was Riccioli a Secret Copernican?" In Borgato, *Giambattista Riccioli e il merito scientifico dei Gesuiti*, 49–77.

Drabkin, I. E. "Aristotle's Wheel: Notes on the History of a Paradox." *Osiris* 9 (1950): 162–98.

Drake, Stillman. *Galileo at Work*. Chicago: University of Chicago Press, 1978.

———. *Galileo's Notes on Motion Arranged in Probable Order of Composition and Presented in Reduced Facsimile*. Florence: Istituto e Museo di Storia della Scienza, 1979.

———. *Galileo Studies: Personality, Tradition, and Revolution*. Ann Arbor: University of Michigan Press, 1970.

———. *History of Free Fall, Aristotle to Galileo*. Toronto: Wall & Thompson, 1989.

Duhem, Pierre Maurice Marie. *Medieval Cosmology: Theories of Infinity, Place, Time, Void, and the Plurality of Worlds*. Translated by Roger Ariew. Chicago: University of Chicago Press, 1985.

Eisenstein, E. *The Printing Press as an Agent of Change: Communications and Cultural Transformations in Early Modern Europe*. 2 vols. Cambridge: Cambridge University Press, 1979.

Elazar, Michael. *Honoré Fabri and the Concept of Impetus: A Bridge between Conceptual Frameworks*. Dordrecht, Netherlands: Springer, 2011.

Elazar, Michael, and Rivka Feldhay. "Honoré Fabri S.J. and Galileo's Law of Fall: What Kind of Controversy?" In Dascal and Boantza, *Controversies within the Scientific Revolution*, 13–32.

Engelberg, Don, and Michael Gertner. "A Marginal Note of Mersenne Concerning the 'Galilean Spiral.'" *Historia Mathematica* 8 (1981): 1–14.

Fabbri, Natacha. "Genesis of Mersenne's 'Harmonie Universelle': The Manuscript 'Livre de La Nature Des Sons.'" *Nuncius* 22, no. 2 (2007): 287–308.

Fabroni, Antonio. *Historiae Academiae Pisanae*. 3 vols. Pisa: Excudebat Cajetanus Mugnainius in aedibus auctoris, 1791–95.

Favaro, Antonio. *Amici e corrispondenti di Galileo*. Edited by Paolo Galluzzi. 3 vols. Florence: Libreria Editrice Salimbeni, 1983.

———. "Gli oppositori di Galileo: I. Antonio Rocco." *Atti del R. Istituto veneto di scienze, lettere ed arti* 50 (1891/92): 1615–36.

———. "Gli oppositori di Galileo. II. Liberto Froidmont." *Atti del R. Istituto veneto di scienze, lettere ed arti* 51 (1892/93): 731–45.

———. "Gli oppositori di Galileo. III. Cristoforo Scheiner." *Atti del R. Istituto veneto di scienze, lettere ed arti* 78, pt. 2 (1919): 1–107.

———. "Oppositori di Galileo: IV. Claudio Berigardo." *Atti del R. Istituto veneto* 79, pt. 2 (1919/20): 39–92.

Feingold, Mordechai. "The Accademia del Cimento and the Royal Society." In Beretta, Clericuzio, and Principe, *Accademia del Cimento and its European Context*, 229–42.

———. "Galileo in England: The First Phase." In Galluzzi, *Novità celesti e crisi del sapere*, 411–20.

———. "The Mathematical Sciences and New Philosophies." In *Seventeenth-Century Oxford*, edited by Nicholas Tyacke, vol. 4 of *The History of the University of Oxford*, 319–448. Oxford: Clarendon, 1997.

———. *The Mathematicians' Apprenticeship: Science, Universities, and Society in England, 1560–1640*. Cambridge: Cambridge University Press, 1984.

———, ed. *The New Science and Jesuit Science: Seventeenth Century Perspectives*. Dordrecht, Netherlands: Kluwer Academic, 2003.

———. "The Origins of the Royal Society Revisited." In *The Practice of Reform in Health, Medicine, and Science, 1500–2000*, edited by Margaret Pelling and Scott Mandelbrote, 167–83. Aldershot, Hampshire: Ashgate, 2005.

Feisenberger, Hellmut Albert, and A. N. L. Munby, eds. *Sale Catalogues of Libraries of Eminent Persons*. Vol. 11, *Scientists*. London: Mansell Information, 1975.

Findlen, Paula. "Living in the Shadow of Galileo: Antonio Baldigiani (1647–1711), a Jesuit Scientist in Late Seventeenth-Century Rome." In *Conflicting Duties: Science, Medicine and Religion in Rome, 1550–1750*, edited by M. P. Donato and J. Kraye, 211–54. London: Warburg Institute, 2009.

Finocchiaro, Maurice A. *Galileo and the Art of Reasoning: Rhetorical Foundations of Logic and Scientific Method*. Dordrecht, Netherlands: D. Reidel, 1980.

———. *Retrying Galileo*. Berkeley: University of California Press, 2005.

Fiocca, Alessandra. "Galileiani e Gesuiti a Ferrara nel Seicento." In Pepe, *Galileo e la scuola galileiana*, 292–309.

Flitner, Andreas. *Erasmus im Urteil seiner Nachwelt; das literarische Erasmus-Bild von Beatus Rhenanus bis zu Jean Le Clerc*. Tübingen: M. Niemeyer, 1952.

Frank, Robert G. *Harvey and the Oxford Physiologists: A Study of Scientific Ideas*. Berkeley: University of California Press, 1980.

Frasca-Spada, M., and N. Jardine. *Books and the Sciences in History*. Cambridge: Cambridge University Press, 2000.

French, Roger. *William Harvey's Natural Philosophy*. Cambridge: Cambridge University Press, 1994.

Gabbey, Alan. "Newton's Mathematical Principles of Natural Philosophy: A Treatise on 'Mechanics'?" In *The Investigation of Difficult Things*, edited by P. M. Harman and A. E. Shapiro, 305–22. Cambridge: Cambridge University Press, 1992.

Galle, Griet, ed. *Peter of Auvergne: Questions on Aristotle's De Caelo—a Critical Edition*. Leuven, Belgium: Leuven University Press, 2003.

Galluzzi, Paolo. "Galileo contro Copernico: Il dibattito sulla prova 'galileiana' di G. B. Riccioli contro il moto della Terra." *Annali dell'Istituto e Museo di Storia della Scienza di Firenze* 2 (1977): 87–148.

———. "Gassendi and l'Affaire Galilée of the Laws of Motion." In *Galileo in Context*, edited by Jürgen Renn, 239–75. Cambridge: Cambridge University Press, 2001.

———. "Gassendi e l'affaire Galilée delle leggi del moto." *Giornale critico della filosofia italiana* 72 (1993): 86–119.

———. *Momento: Studi Galileiani*. Rome: Edizioni dell'Ateneo & Bizzarri, 1979.

———, ed. *Novità celesti e crisi del sapere; Atti del Convegno Internazionale di Studi Galileiani*. Florence: Giunti Barbera, 1984.

———. "La scienza davanti alla Chiesa e al Principe in una polemica universitaria del secondo Seicento." In *Studi in onore di Arnaldo d'Addario*, edited by Luigi Borgia, 4, pt. 1: 1317–44. 4 vols. Lecce: Conte, 1995.

———. "The Sepulchers of Galileo: The 'Living' Remains of a Hero of Science." In *The Cambridge Companion to Galileo*, edited by Peter Machamer, 417–47. Cambridge: Cambridge University Press, 1998.

Garber, D. "On the Frontlines of the Scientific Revolution: How Mersenne Learned to Love Galileo." *Perspectives on Science* 12, no. 2 (2004): 135–60.

Garcia, Stéphane. *Élie Diodati et Galilée: Naissance d'un réseau scientifique dans l'Europe du XVIIe siècle*. Florence: Olschki, 2004.

Gardair, J. M. "Elia Diodati e la diffusione europa del *Dialogo*." In Galluzzi, *Novità celesti e crisi del sapere*, 391–98.

Gaukroger, S. *Explanatory Structures: A Study of Concepts of Explanation in Early Physics and Philosophy*. Atlantic Highlands, NJ: Humanities Press, 1978.

Gavagna, Veronica. "I Gesuiti e la polemica sul vuoto: Il contributo di Paolo Casati." In *Gesuiti e università in Europa*, edited by Gian Paolo Brizzi and Roberto Greci, 325–38. Bologna: Cooperativa Libraria Universitaria Editrice Bologna, 2002.

———. "Paolo Casati e la scuola galileiana." In Pepe, *Galileo e la scuola galileiana*, 311–26.

Gingerich, Owen. *An Annotated Census of Copernicus' De Revolutionibus (Nuremberg, 1543 and Basel, 1566)*. Leiden: Brill, 2002.

———. *The Book Nobody Read: Chasing the Revolutions of Nicolaus Copernicus*. New York: Walker, 2004.

Gingerich, Owen, and Robert S. Westman. "The Wittich Connection: Conflict and Priority in Late Sixteenth-Century Cosmology." *Transactions of the American Philosophical Society* 78, no. 7 (1988): i–viii, 1–148.

Giusti, Enrico. *Euclides reformatus: La teoria delle proporzioni nella scuola galileiana*. Turin: Bollati Boringhieri, 1993.

———. "Galileo all'origine delle ricerche della scuola galileiana." In Pepe, *Galileo e la scuola galileiana*, 7–22.

———. "A Master and His Pupils: Theories of Motion in the Galilean School." In Palmerino and Thijssen, *Reception of the Galilean Science of Motion*, 119–35.

Gomez Lopez, Susana. "Donato Rossetti et le Cercle pisan." In *Géométrie, atomisme et vide dans l'école de Galilée*, edited by Egidio Festa, Vincent Jullien, and Maurizio Torrini, 281–97. Fontenay /Saint-Cloud: ENS / Istituto e Museo di Storia della Scienza, 1999.

———. "Dopo Borelli: La scuola galileiana a Pisa." In Pepe, *Galileo e la scuola galileiana*, 223–32.

———. *Le passioni degli atomi: Montanari e Rossetti; Una polemica tra galileiani*. Florence: Olschki, 1997.

Gorman, Michael John. "Jesuit Explorations of the Torricellian Space: Carp-bladders and Sulphurous Fumes." *Mélanges de l'Ecole française de Rome: Italie et Méditerranée* 106, no. 1 (1994): 7–32.

Goulding, Robert. "Polemic in the Margin: Henry Savile and Joseph Scaliger on Squaring the Circle." In *Scientia in margine : Études sur les marginalia dans les manuscrits scientifiques du Moyen Âge à la Renaissance*, edited by D. Jacquart and C. Burnett, 241–59. Geneva: Droz, 2005.

Grafton, Anthony. *The Culture of Correction in Renaissance Europe*. London: British Library, 2011.

———. *Defenders of the Text: The Traditions of Scholarship in an Age of Science, 1450–1800*. Cambridge, MA: Harvard University Press, 1991.

———. "The Humanist as Reader." In *A History of Reading in the West*, edited by Guglielmo Cavallo, Roger Chartier, and Lydia G. Cochrane, 179–212. Amherst: University of Massachusetts Press, 1999.

———. "The Republic of Letters in the American Colonies: Francis Daniel Pastorius Makes a Notebook." *American Historical Review* 117, no. 1 (2012): 1–39.

———. "Teacher, Text and Classroom: A Study from a Parisian College." *History of Universities* 9 (1990): 73–118.

Grafton, Anthony, and Lisa Jardine. *From Humanism to the Humanities: Education and Liberal Arts in Fifteenth- and Sixteenth-Century Europe*. London: Duckworth, 1986.

Grafton, Anthony, and Joanna Weinberg. *"I Have Always Loved the Holy Tongue": Isaac Casaubon, the Jews, and a Forgotten Chapter in Renaissance Scholarship*. Cambridge, MA: Belknap Press of Harvard University Press, 2011.

Grant, Edward. *Much Ado about Nothing: Theories of Space and Vacuum from the Middle Ages to the Scientific Revolution*. Cambridge: Cambridge University Press, 1981.

Grendler, Paul F. *The Roman Inquisition and the Venetian Press*. Princeton, NJ: Princeton University Press, 1977.

———. *Schooling in Renaissance Italy: Literacy and Learning, 1300–1600*. Baltimore: Johns Hopkins University Press, 1991.

———. *The Universities of the Italian Renaissance*. Baltimore: Johns Hopkins University Press, 2002.

Groote, Inga Mai, Bernhard Kölbl, and Susan Forscher Weiss. "Evidence for Glarean's Music Lectures from His Students' Books: Congruent Annotations in the *Epitome* and the *Dodekachordon*." In *Heinrich Glarean's Books: The Intellectual World of a Sixteenth-Century Musical Humanist*, edited by Iain Fenlon and Inga Mai Groote, 280–302. Cambridge: Cambridge University Press, 2013.

Guicciardini, Niccolò. *Reading the* Principia: *The Debate on Newton's Mathematical Methods for Natural Philosophy from 1687 to 1736*. Cambridge: Cambridge University Press, 1999.

Hall, A. Rupert. *The Scientific Revolution, 1500–1800*. Boston: Beacon, 1966.

Hall, Crystal. "Galileo's Library Reconsidered." *Galilaeana* 12 (2015): 29–82.

——. *Galileo's Reading*. Cambridge: Cambridge University Press, 2014.

Hankins, James. *Plato in the Italian Renaissance*. 2 vols. Leiden: Brill, 1990.

Hankins, Thomas L. *Science and the Enlightenment*. Cambridge: Cambridge University Press, 2010.

Harrison, John R. *The Library of Isaac Newton*. Cambridge: Cambridge University Press, 1978.

Heesen, Anke Te. "The Notebook: A Paper Technology." In *Making Things Public: Atmospheres of Democracy*, edited by Bruno Latour and Peter Weibel, 582–89. Cambridge, MA: MIT Press, 2005.

Heilbron, John L. *Electricity in the 17th and 18th Centuries: A Study of Early Modern Physics*. Berkeley: University of California Press, 1979.

——. *Galileo*. Oxford: Oxford University Press, 2010.

——. *The Sun in the Church: Cathedrals as Solar Observatories*. Cambridge, MA: Harvard University Press, 1999.

Helbing, Mario. *La filosofia di Francesco Buonamici, professore di Galileo a Pisa*. Pisa: Nistri-Lischi, 1989.

Hellyer, Marcus. "'Because the Authority of My Superiors Commands': Censorship, Physics and the German Jesuits." *Early Science and Medicine* 1, no. 3 (1996): 319–54.

——. *Catholic Physics: Jesuit Natural Philosophy in Early Modern Germany*. Notre Dame, IN: University of Notre Dame Press, 2005.

Hess, G. "Fundamenta fürstlicher Tugend: Zum Stellenwert der Sentenz im Rahmen der voruniversitären Ausbildung Herzog August d. J." In *Sammeln, Ordnen, Veranschaulichen: Zur Wissenskompilatorik in der Frühen Neuzeit*, edited by Frank Büttner, Markus Friedrich, and Helmut Zedelmaier, 131–74. Münster: LIT Verlag, 2003.

Hess, Volker, and J. Andrew Mendelsohn. "Case and Series: Medical Knowledge and Paper Technology, 1600–1900." *History of Science* 48 (2010): 287–314.

Hunt, R. W., and A. G. Watson. *Bodleian Library Quarto Catalogues. IX. Digby Manuscripts*. Oxford: Bodleian Library, 1999.

Hunter, Michael. *The Boyle Papers: Understanding the Manuscripts of Robert Boyle*. Aldershot, Hampshire: Ashgate, 2007.

——. *Establishing the New Science: The Experience of the Early Royal Society*. Woodbridge, Suffolk: Boydell, 1989.

Iofrida, Manlio. "La filosofia e la medicina (1543–1737)." In *Storia dell'Università di Pisa*, edited by E. Amatori, 1:289–338. Pisa: Pacini, 1993.

Jaki, Stanley L. Introduction to *To Save the Phenomena: An Essay on the Idea of Physical Theory from Plato to Galileo*, by Pierre Duhem. Chicago: University of Chicago Press, 1985.

Jardine, Lisa. "Monuments and Microscopes: Scientific Thinking on a Grand Scale in the Early Royal Society." *Notes and Records of the Royal Society of London* 55, no. 2 (2001): 289–308.

——. *On a Grander Scale: The Outstanding Career of Sir Christopher Wren*. London: HarperCollins, 2002.

Jardine, Lisa, and Anthony Grafton. "'Studied for Action': How Gabriel Harvey Read His Livy." *Past and Present* 129, no. 1 (1990): 30–78.

Jardine, Nicholas. "Demonstration, Dialectic, and Rhetoric in Galileo's *Dialogue*." In *The Shapes of Knowledge from the Renaissance to the Enlightenment*, edited by R. Kelley and R. H. Popkin, 101–21. Dordrecht, Netherlands: Kluwer, 1991.

Johns, Adrian. *The Nature of the Book: Print and Knowledge in the Making*. Chicago: University of Chicago Press, 1998.

———. "Reading and Experiment in the Early Royal Society." In *Reading, Society and Politics in Early Modern England*, edited by Kevin Sharpe and Steven N. Zwicker, 244–71. Cambridge: Cambridge University Press, 2003.

Jolley, Nicholas. "The Reception of Descartes' Philosophy." In *The Cambridge Companion to Descartes*, edited by John Cottingham, 393–423. Cambridge: Cambridge University Press, 2006.

Keller, Vera. "Accounting for Invention: Guido Pancirolli's Lost and Found Things and the Development of *Desiderata*." *Journal of the History of Ideas* 73, no. 2 (2012): 223–45.

———. *Knowledge and the Public Interest, 1575–1725*. New York: Cambridge University Press, 2015.

———. "The 'New World of Sciences': The Temporality of the Research Agenda and the Unending Ambitions of Science." *Isis* 103, no. 4 (2012): 727–34.

Kors, A. C. *The Orthodox Sources of Disbelief*. Vol. 1 of *Atheism in France, 1650–1729*. Princeton, NJ: Princeton University Press, 1990.

Koyré, Alexandre. "A Documentary History of the Problem of Fall from Kepler to Newton." *Transactions of the American Philosophical Society* 45 (1955): 329–95.

———. "An Experiment in Measurement." *Proceedings of the American Philosophical Society* 97 (1953): 222–37.

———. *Galileo Studies*, translated by J. Mepham. Atlantic Highlands, NJ: Humanities Press, 1978.

Kretzmann, Norman. "Syncategoremata, Exponibilia, Sophismata." In *The Cambridge History of Later Medieval Philosophy*, edited by Norman Kretzmann, Anthony Kenny, and Jan Pinborg, 211–45. Cambridge: Cambridge University Press, 1982.

Kusukawa, Sachiko. *Picturing the Book of Nature: Image, Text, and Argument in Sixteenth-Century Human Anatomy and Medical Botany*. Chicago : University of Chicago Press, 2012.

Lattis, J. M. *Between Copernicus and Galileo: Christoph Clavius and the Collapse of Ptolemaic Cosmology*. Chicago: University of Chicago Press, 1994.

Lenoble, Robert. *Mersenne, ou, La naissance du mécanisme*. Paris: J. Vrin, 1971.

Leong, Elaine. "'Herbals She Peruseth': Reading Medicine in Early Modern England." *Renaissance Studies* 28, no. 4 (2014): 556–78.

Lewis, John. "Mersenne as Translator and Interpreter of the Works of Galileo." *MLN* 127, no. 4 (2012): 754–82.

Lewis, Rhodri. *Language, Mind, and Nature: Artificial Languages in England from Bacon to Locke*. Cambridge: Cambridge University Press, 2007.

Lindberg, D. C. "Conceptions of the Scientific Revolution from Bacon to Butterfield: A Preliminary Sketch." In *Reappraisals of the Scientific Revolution*, edited by D. C. Lindberg and R. S. Westman, 1–26. Cambridge: Cambridge University Press, 1990.

Lines, David A. "Teaching Physics in Louvain and Bologna." In *Scholarly Knowledge : Textbooks in Early Modern Europe*, edited by Emidio Campi, 183–203. Geneva: Droz, 2008.

Malcolm, Noel. "Hobbes and the Royal Society." In *Aspects of Hobbes*, 317–35. Oxford: Clarendon, 2002.

Mandelbrote, Scott. *Footprints of the Lion: Isaac Newton at Work*. Cambridge: Cambridge University Press, 2001.

Mansfield, Bruce. *Phoenix of His Age: Interpretations of Erasmus c. 1550–1750*. Toronto: University of Toronto Press, 1979.

Marenbon, John. *Later Medieval Philosophy (1150–1350)*. London: Routledge, 1987.

Marr, Alexander. *Between Raphael and Galileo: Mutio Oddi and the Mathematical Culture of Late Renaissance Italy*. Chicago: University of Chicago Press, 2011.

Martin, Craig. *Renaissance Meteorology: Pomponazzi to Descartes*. Baltimore: Johns Hopkins University Press, 2011.

Middleton, W. E. K. *The Experimenters: A Study of the Accademia del Cimento*. Baltimore: Johns Hopkins Press, 1971.

Miller, Peter N. *Peiresc's Europe: Learning and Virtue in the Seventeenth Century*. New Haven, CT: Yale University Press, 2000.

Mir, Gabriel Codina. *Aux sources de la pédagogie des Jésuites: Le "Modus parisiensis."* Rome: Institutum Historicum Societatis Iesu, 1968.

Montacutelli, Stefania. "Air 'Particulae' and Mechanical Motions: From the Experiments of the Cimento Academy to Borelli's Hypotheses on the Nature of Air." In Beretta, Clericuzio, and Principe, *Accademia del Cimento and Its European Context*, 59–72.

Moss, Ann. *Printed Commonplace-Books and the Structuring of Renaissance Thought*. Oxford: Clarendon, 1996.

Moss, Jean Dietz. *Novelties in the Heavens: Rhetoric and Science in the Copernican Controversy*. Chicago: University of Chicago Press, 1993.

Müller-Wille, Staffan, and Sara Scharf. "Indexing Nature: Carl Linnaeus (1707–1778) and His Fact-Gathering Strategies." Edited by Jon Adams. Working Papers on the Nature of Evidence: How Well Do Facts Travel? 36/08. 2009.

Murdoch, J. E. "Infinity and Continuity." In *The Cambridge History of Later Medieval Philosophy*, edited by Norman Kretzmann, Anthony Kenny, and Jan Pinborg, 564–91. Cambridge: Cambridge University Press, 1982.

Navarro Brotons, Victor. "Filosofia natural y disciplinas matematicas en la España del siglo XVII." In Pepe, *Galileo e la scuola galileiana*, 89–106.

Neagu, Cristina. "Time Capsule under Restoration: The Allestree Library." *Christ Church Library Newsletter* 7, no. 2 (Hilary 2011): 15–17.

Needham, Paul. *Galileo Makes a Book: The First Edition of Sidereus Nuncius, Venice 1610*. Vol. 2 of *Galileo's O*, edited by Horst Bredekamp. Berlin: Akademie Verlag, 2011.

Nelles, Paul. "Libros de Papel, Libri Bianchi, Libri Papyracei. Note-Taking Techniques and the Role of Student Notebooks in the Early Jesuit Colleges." *Archivum Historicum Societatis Iesu* 76 (2007): 75–111.

——. "Seeing and Writing: The Art of Observation in the Early Jesuit Missions." *Intellectual History Review* 20, no. 3 (2010): 317–33.

Newman, William R., and Lawrence M. Principe. *Alchemy Tried in the Fire: Starkey, Boyle, and the Fate of Helmontian Chymistry.* Chicago: University of Chicago Press, 2002.

Nuovo, Angela. *The Book Trade in the Italian Renaissance.* Boston: Brill, 2013.

O'Malley, John W., Gauvin A. Bailey, Steven J. Harris, and T. Frank Kennedy, eds. *The Jesuits: Cultures, Sciences, and the Arts, 1540–1773.* Toronto: University of Toronto Press, 1999.

——, eds. *The Jesuits II: Cultures, Sciences, and the Arts, 1540–1773.* Toronto: University of Toronto Press, 2006.

Oosterhoff, Richard J. "A Book, a Pen, and the Sphere: Reading Sacrobosco in the Renaissance." *History of Universities* 28, no. 2 (2015): 1–54.

Palmerino, Carla Rita. "Galileo's and Gassendi's Solutions to the Rota Aristotelis Paradox: A Bridge between Matter and Motion Theories." In *Late Medieval and Early Modern Corpuscular Matter Theories*, edited by Christoph H. Lüthy, John E. Murdoch, and William R. Newman, 381–422. Leiden: Brill, 2001.

——. "Galileo's Theories of Free Fall and Projectile Motion as Interpreted by Pierre Gassendi." In Palmerino and Thijssen, *Reception of the Galilean Science of Motion*, 137–64.

——. "Gassendi's Reinterpretation of the Galilean Theory of Tides." *Perspectives on Science* 12, no. 2 (2004): 212–37.

——. "Infinite Degrees of Speed: Marin Mersenne and the Debate over Galileo's Law of Free Fall." *Early Science and Medicine* 4 (1999): 269–328.

——. "La fortuna della scienza galileiana nelle Province Unite." In Pepe, *Galileo e la scuola galileiana*, 61–79.

——. "Two Jesuit Responses to Galileo's Science of Motion: Honoré Fabri and Pierre Le Cazre." In Feingold, *New Science and Jesuit Science*, 187–228.

——. "Una nuova scienza della materia per la scienza nuova del moto: La discussione dei paradossi dell'infinito nella Prima Giornata dei *Discorsi* galileiani." In *Atomismo e continuo nel XVII secolo*, edited by Egidio Festa and Romano Gatto, 275–319. Naples: Vivarium, 2000.

Palmerino, Carla Rita, and J. M. M. H. Thijssen, eds. *The Reception of the Galilean Science of Motion in Seventeenth-Century Europe.* Dordrecht, Netherlands: Kluwer Academic, 2004.

Palmieri, Paolo. "A Phenomenology of Galileo's Experiments with Pendulums." *British Journal for the History of Science* 42, no. 4 (2009): 479–513.

——. *Reenacting Galileo's Experiments: Rediscovering the Techniques of Seventeenth-Century Science.* Lewiston, NY: Edwin Mellen, 2008.

Parenty, Hélène. *Isaac Casaubon, helléniste: Des studia humanitatis à la philologie.* Geneva: Droz, 2009.

Park, Katharine. "Observation in the Margins, 500–1500." In *Histories of Scientific Observation*, edited by Lorraine Daston and Elizabeth Lunbeck, 15–44. Chicago: University of Chicago Press, 2011.

Park, Katharine, and Lorraine Daston, eds. *Early Modern Science.* Vol. 3 of *The Cambridge History of Science.* Cambridge: Cambridge University Press, 2006.

Pepe, Luigi, ed. *Galileo e la scuola galileiana nelle università del Seicento.* Bologna: Clueb, 2011.

Philip, Ian. *The Bodleian Library in the Seventeenth and Eighteenth Centuries*. Oxford: Clarendon, 1983.

Picciotto, Joanna. *Labors of Innocence in Early Modern England*. Cambridge, MA: Harvard University Press, 2010.

——. "Scientific Investigations: Experimentalism and Paradisal Return." In *A Concise Companion to the Restoration and Eighteenth Century*, edited by Cynthia Wall, 36–57. Malden, MA: Blackwell, 2005.

Pomata, Gianna. "Observation Rising: Birth of an Epistemic Genre, 1500–1650." In *Histories of Scientific Observation*, edited by Lorraine Daston and Elizabeth Lunbeck, 45–80. Chicago: University of Chicago Press, 2011.

Poole, William. *The World Makers: Scientists of the Restoration and the Search for the Origins of the Earth*. Oxford: Peter Lang, 2010.

Poole, William, Felicity Henderson, and Yelda Nasifoglu. "Hooke's Books Database / Robert Hooke's Books." *Robert Hooke's Books*. Accessed 21 September 2015. http://www.hookesbooks.com/hookes-books-database/.

Porter, R. "The Scientific Revolution and Universities." In *Universities in Early Modern Europe (1500–1800)*, edited by H. D. Ridder-Symoens, vol. 2 of *A History of the University in Europe*, 531–62. Cambridge: Cambridge University Press, 1996.

Preti, Cesare. "Giannetti, Pascasio." In *Dizionario biografico degli italiani*, vol. 54, edited by Mario Caravale and Giuseppe Pignatelli. Rome: Istituto dell'Enciclopedia Italiana, 2000.

Principe, Lawrence M. *The Aspiring Adept: Robert Boyle and His Alchemical Quest*. Princeton, NJ: Princeton University Press, 1998.

Procissi, Angiolo. *La collezione Galileiana della Biblioteca nazionale di Firenze*. 5 parts in 3 vols. Rome: Istituto Poligrafico dello Stato, 1959–94.

Raphael, Renée. "Copernicanism in the Classroom: Jesuit Natural Philosophy and Mathematics after 1633." *Journal for the History of Astronomy* 46, no. 4 (2015): 419–40.

——. "Galileo's *Discorsi* and Mersenne's *Nouvelles Pensées*: Mersenne as a Reader of Galilean 'Experience.'" *Nuncius* 23, no. 1 (2008): 7–36.

——. "Galileo's *Discorsi* as a Tool for the Analytical Art." *Annals of Science* 72, no. 1 (2014): 99–123.

——. "Galileo's *Two New Sciences* as a Model of Reading Practice." *Journal of the History of Ideas*, forthcoming.

——. "Making Sense of Day 1 of the *Two New Sciences*: Galileo's Aristotelian-Inspired Agenda and His Jesuit Readers." *Studies in History and Philosophy of Science* 42 (2011): 479–91.

——. "Printing Galileo's *Discorsi*: A Collaborative Affair." *Annals of Science* 69, no. 4 (2012): 485–513.

——. "Reading Galileo's *Discorsi* in the Early Modern University." *Renaissance Quarterly* 68, no. 2 (2015): 558–96.

——. "Teaching through Diagrams: Galileo's *Dialogo* and *Discorsi* and His Pisan Readers." *Early Science and Medicine* 18, nos. 1–2 (2013): 201–30.

Renn, Jürgen. "Galileo's Manuscripts on Mechanics: The Project of an Edition with Full Critical Apparatus of Mss. Gal. Codex 72." *Nuncius* 3, no. 1 (1988): 193–241.

Renn, Jürgen, and Peter Damerow. *The Equilibrium Controversy: Guidobaldo del Monte's Critical Notes on the Mechanics of Jordanus and Benedetti and Their Historical and Conceptual Background*. Berlin: Edition Open Access, 2012.

Renn, Jürgen, Peter Damerow, and Simone Rieger. "Hunting the White Elephant: When and How Did Galileo Discover the Law of Fall?" In *Galileo in Context*, edited by Jürgen Renn, 29–149. Cambridge: Cambridge University Press, 2001.

Robinson, H. W. "An Unpublished Letter of Dr Seth Ward Relating to the Early Meetings of the Oxford Philosophical Society." *Notes and Records of the Royal Society of London* 7, no. 1 (1 December 1949): 68–70.

Rostenberg, Leona. *The Library of Robert Hooke: The Scientific Book Trade of Restoration England*. Santa Monica, CA: Modoc, 1989.

Roux, Sophie. "An Empire Divided: French Natural Philosophy (1670–1690)." In *The Mechanization of Natural Philosophy*, edited by Daniel Garber and Sophie Roux, 55–95. Dordrecht, Netherlands: Springer, 2013.

Saenger, Paul Henry. *Space between Words: The Origins of Silent Reading*. Stanford, CA: Stanford University Press, 1997.

Schemmel, Matthias. *The English Galileo : Thomas Harriot's Work on Motion as an Example of Preclassical Mechanics*. 2 vols. London: Springer, 2008.

Schmitt, Charles B. "Eclectic Aristotelianism." In *Aristotle and the Renaissance*, 89–109. Cambridge, MA: Harvard University Press, 1983.

——. "Experience and Experiment: A Comparison of Zabarella's View with Galileo's in *De Motu*." *Studies in the Renaissance* 16 (1969): 80–138.

——. "The Faculty of Arts at Pisa at the Time of Galileo." In *Studies in Renaissance Philosophy and Science*, 243–72. London: Variorum Reprints, 1981.

——. "Galilei and the Seventeenth-Century Text-Book Tradition." In *Reappraisals in Renaissance Thought*, edited by C. Webster, 217–28. London: Variorum Reprints, 1989.

——. "Renaissance Aristotelianisms." In *Aristotle and the Renaissance*, 10–33. Cambridge, MA: Harvard University Press, 1983.

Settle, Thomas B. "An Experiment in the History of Science." *Science* 133 (1961): 19–23.

——. "Galileo and Early Experimentation." In *Springs of Scientific Creativity*, edited by H. T. D. R. Aris. Minneapolis: University of Minnesota Press, 1983.

——. "Galileo's Use of Experiment as a Tool of Investigation." In *Galileo, Man of Science*, edited by E. McMullin, 315–37. New York: Basic Books, 1968.

Shapin, Steven. *The Scientific Revolution*. Chicago: University of Chicago Press, 1998.

——. *A Social History of Truth: Civility and Science in Seventeenth-Century England*. Chicago: University of Chicago Press, 1994.

Shapin, Steven, and Simon Schaffer. *Leviathan and the Air-Pump: Hobbes, Boyle, and the Experimental Life*. Princeton, NJ: Princeton University Press, 1985.

Shea, William R. *Designing Experiments and Games of Chance*. Canton, MA: Science History, 2003.

——. *Galileo's Intellectual Revolution: Middle Period, 1610–1632*. New York: Science History, 1977.

——. "Marin Mersenne: Galileo's 'Traduttore-Traditore.'" *Annali dell'Istituto e Museo di Storia della Scienza di Firenze* 2 (1977): 55–70.

Sherman, William H. *John Dee: The Politics of Reading and Writing in the English Renaissance*. Amherst: University of Massachusetts Press, 1995.

———. *Used Books: Marking Readers in Renaissance England*. Philadelphia: University of Pennsylvania Press, 2008.

Siebert, H. "Kircher and His Critics: Censorial Practice and Pragmatic Disregard in the Society of Jesus." In *Athanasius Kircher: The Last Man Who Knew Everything*, edited by P. Findlen, 79–104. New York: Routledge, 2004.

Sobel, D. *Galileo's Daughter: A Historical Memoir of Science, Faith, and Love*. New York: Walker, 1999.

Soll, Jacob. "Amelot de La Houssaye (1634–1706) Annotates Tacitus." *Journal of the History of Ideas* 61 (2000): 167–87.

Sommervogel, Carlos, ed. *Bibliothèque de la Compagnie de Jésus*. 12 vols. Louvain, Belgium: Editions de la Bibliothèque S.J., 1960.

Stabile, Giorgio. *Claudio Berigard, 1592–1663: Contributo alla storia dell'atomismo seicentesco*. Rome: Istituto di filosofia dell'Università, 1975.

———. "Il primo oppositore del *Dialogo*: Claude Bérigard." In Galluzzi, *Novità celesti e crisi del sapere*, 277–82.

Stedall, Jacqueline A. *The Arithmetic of Infinitesimals: John Wallis 1656*. New York: Springer, 2004.

———. *A Discourse concerning Algebra: English Algebra to 1685*. Oxford: Oxford University Press, 2002.

Stolzenberg, D. "Oedipus Censored: Censurae of Athanasius Kircher's Works in the Archivum Romanum Societatis Iesu." *Archivum Historicum Societatis Iesu* 73, no. 145 (2004): 3–52.

Taton, René. *Les origines de l'Académie royale des sciences*. Paris: Palais de la découverte, 1966.

Valleriani, Matteo. *Galileo Engineer*. London: Springer, 2010.

———. "A View on Galileo's 'Ricordi Autografi': Galileo Practitioner in Padua." In *Largo campo di filosofare*, edited by José Montesinos and Carlos Solís, 281–91. La Orotava, Tenerife: Fundación Canaria Orotava de Historia de la Ciencia, 2001.

van Helden, Albert. "Longitude and the Satellites of Jupiter." In *The Quest for Longitude*, edited by W. J. H. Andrewes, 85–100. Cambridge, MA: Harvard Collection of Historical Scientific Instruments, 1996.

Vilain, Christiane. "Christiaan Huygens' Galilean Mechanics." In Palmerino and Thijssen, *Reception of the Galilean Science of Motion*, 185–98.

Villoslada, Ricardo García. *Storia del Collegio Romano dal suo inizio (1551) alla soppressione della Compagnia di Gesù (1773)*. Rome: Aedes Universitatis Gregorianae, 1954.

Wallace, William A. *Galileo and His Sources: The Heritage of the Collegio Romano in Galileo's Science*. Princeton, NJ: Princeton University Press, 1984.

Waquet, Françoise. *Le modèle français et l'Italie savante: Conscience de soi et perception de l'autre dans la République des Lettres (1660–1750)*. Rome: École Francaise de Rome, 1989.

Watson, William Patrick. *Science, Medicine, Natural History, Catalogue 19*. London: William Patrick Watson Antiquarian Books, 2013.

Webster, Charles. *The Great Instauration: Science, Medicine and Reform, 1626–1660*. London: Duckworth, 1975.

Westfall, Richard S. *Force in Newton's Physics: The Science of Dynamics in the Seventeenth Century*. London: Macdonald, 1971.

Westman, Robert S. *The Copernican Question: Prognostication, Skepticism, and Celestial Order*. Berkeley: University of California Press, 2011.

———. "The Reception of Galileo's 'Dialogue': A Partial World Census of Extant Copies." In Galluzzi, *Novità celesti e crisi del sapere*, 329–71.

Wilding, Nick. *Galileo's Idol: Gianfrancesco Sagredo and the Politics of Knowledge*. Chicago: University of Chicago Press, 2014.

———. "Manuscripts in Motion: The Diffusion of Galilean Copernicanism." *Italian Studies* 66, no. 2 (2011): 221–33.

———. "The Return of Thomas Salusbury's *Life of Galileo* (1664)." *British Journal for the History of Science* 41, no. 2 (June 2008): 241–65.

Wisan, Winifred Lovell. "Galileo's 'De systemate mundi' and the new Mechanics." In Galluzzi, *Novità celesti e crisi del sapere*, 41–47.

———. "Galileo's Scientific Method: A Reexamination." In *New Perspectives on Galileo*, edited by Robert E. Butts and Joseph C. Pitt, 1–57. Dordrecht, Netherlands: D. Reidel, 1978.

———. "The New Science of Motion: A Study of Galileo's 'De Motu Locali.'" *Archive for History of Exact Sciences* 13 (1974): 103–306.

———. "On the Chronology of Galileo's Writings." *Annali dell'Istituto e Museo di Storia della Scienza di Firenze* 9 (1984): 85–88.

Wootton, David. *Galileo: Watcher of the Skies*. New Haven, CT: Yale University Press, 2010.

Yale, Elizabeth. "Marginalia, Commonplaces, and Correspondence: Scribal Exchange in Early Modern Science." *Studies in History and Philosophy of Biological and Biomedical Sciences* 42 (2011): 193–202.

———. *Sociable Knowledge: Natural History and the Nation in Early Modern Britain*. Philadelphia: University of Pennsylvania Press, 2016.

Yeo, Richard. "Between Memory and Paperbooks: Baconianism and Natural History in Seventeenth-Century England." *History of Science* 45 (2007): 1–46.

———. "John Locke's 'New Method' of Commonplacing: Managing Memory and Information." *Eighteenth-Century Thought* 2 (2004): 1–38.

———. "Loose Notes and Capacious Memory: Robert Boyle's Note-Taking and Its Rationale." *Intellectual History Review* 20, no. 3 (2010): 335–54.

———. *Notebooks, English Virtuosi, and Early Modern Science*. Chicago: University of Chicago Press, 2014.

Zanfredini, M. "Mauro, Silvestro." In *Diccionario histórico de la Compañía de Jesús: Biográfico-temático*, edited by Charles E. O'Neill and Joaquín M. Dominguez, 3:2583. 4 vols. Madrid: Universidad Pontificia Comillas, 2001.

Zehnacker, Françoise. "Bibliothèque Sainte-Geneviève." In *Patrimoine des bibliothèques de France, un guide des régions*, project director Anne-Marie Reder, 1:252–61. 11 vols. Paris: Payot, 1995.